BIO-DIESEL : Jatropha Curcas

(A PROMISING SOURCE)

BIO-DIESEL
Jatropha Curcas

(A PROMISING SOURCE)

By

Dr. Mohammad Arif

Dr. Zakwan Ahmed

Govt. of India, Ministry of Defence

Defence Research & Development Organization

Defence Institute of Bio-Energy Research

Haldwani (Uttarakhand) - 263 139

2009

SSPH

SATISH SERIAL PUBLISHING HOUSE

403, Express Tower, Commercial Complex, Azadpur, Delhi-110033 (India)

Phone : 011-27672852, Fax : 91-11-27672046

E-mail : info@satishserial.com, hkjain1975@yahoo.com

Website : www.satishserial.com

Published by :

SATISH SERIAL PUBLISHING HOUSE
403, Express Tower, Commercial Complex, Azadpur, Delhi-110033 (INDIA)
Phone : 011-27672852 Fax : 91-11-27672046
E-mail : info@satishserial.com, hkjain1975@yahoo.com

ISBN 81-89304-65-8

Typeset at: Laxmi Art Creations # 9811482328
Printed at: Salasar Imaging Systems, Delhi-35

Dedicated to

M Natarajan
Scientific Adviser to *Raksha Mantri*
Secretary & DG R&D (DRDO)

"The use of vegetable oils for engine fuel may seem insignificant today. But such oils may become in course of time as important as petroleum and the coal tar products of the present time"

Rudolf Diesel, 1912

"Oil is of vast importance in the world today. A country that does not produce its own oil is in the weak position. From the point of view of Defence, the absence of oil is a fatal weakness"

26 May 1956

Late PANDIT JAWAHAR LAL NEHRU
The first Prime Minister of India

ADDRESS TO THE NATION BY PRESIDENT OF INDIA

"In present scenario of dependability on fossil material based system with its uncertainty, it is essential that an energy policy is evolved with new energy avenues. Bio-diesel has the potential to transform the oil sector vision to produce 60 MMT Bio-diesel by 2030".

9 June, 2006

DR. A P J ABDUL KALAM
Bharat Ratna

MASSIVE WORK ON JATROPHA- 2003

Machineries are run by Diesel of Fossil Fuel Origin

causes

Emission of Co_2 (Pollution)

↓

China & India - Rank Top 5 Co_2 Emitter

&

India after China, Japan, Russia and US

(5th Largest Importer of Petroleum)

↓

Dependent on External Agency Sources

(Highly Unstable Region)

↓

ENERGY SECURITY IS NATIONAL KEY ISSUE

सत्यमेव जयते
दीपक गुप्ता
Deepak Gupta

सचिव
भारत सरकार
नवीन और नवीकरणीय ऊर्जा मंत्रालय
SECRETARY
GOVERNMENT OF INDIA
MINISTRY OF NEW AND RENEWABLE ENERGY

13.3.2009

MESSAGE

Petroleum resources are finite and, therefore, search for alternative is continued all over the world, particularly in India. India's energy security would remain vulnerable until alternative fuels to substitute / supplement petro-based fuels are developed based on indigenously produced renewable feedstocks. In biofuels, the country has a ray of hope in providing energy security. Biofuels are environment friendly fuels and their utiliz ation would address global concerns about containment of carbon emissions. The transportation sector has been identified as a major polluting sector. Use of biofuels have, therefore, become compelling in view of the tightening automotive vehicle emission standards to curb air pollution.

In the Indian context, bio-diesel is of special importance as the Indian economy is more dependent on diesel than petrol. Bio-diesel is derived from vegetable oils and, therefore, provide a strategic advantage to promote sustainable development and supplement conventional diesel fuel in meeting the rapidly increasing requirements for transportation fuels associated with high economic growth, as well as in meeting the energy needs of India's vast rural population. Biofuels can increasingly satisfy these energy needs in an environmentally benign and cost-effective manner while reducing dependence on import of fossil fuels and thereby providing a higher degree of National Energy Security.

The book by Dr. Mohommad Arif and Dr. Zakwan Ahmed is lucidly written and outlines aspects relating to Agro-technology of Jatropha curcas; products processing; their utility; and future prospects of bio-diesel. I congratulate both the authors for their efforts in compilation of this book, which will be of immense use for the scientists and other stakeholders engaged in the production or bio-diesel.

(DEEPAK GUPTA)

ब्लॉक नं 14, केन्द्रीय कार्यालय परिसर, लोदी रोड, नई दिल्ली - 110003
Block No. 14, CGO Complex, Lodi Road, New Delhi - 110 003
Tel. : 011-24361481, 24362772 • Fax : 011-24367329 • E-mail : secymnes@nic.in

सत्यमेव जयते

डॉ० डब्ल्यू. सेल्वामूर्ति
विशिष्ट वैज्ञानिक एवं मुख्य नियंत्रक अनुसंधान तथा विकास
Dr. W. Selvamurthy
Distinguished Scientist & Chief Controller Research & Development

रक्षा मंत्रालय, भारत सरकार
रक्षा अनुसंधान तथा विकास संगठन
201, डी आर डी ओ भवन
नई दिल्ली–110105
Government of India
Ministry of Defence
Defence Research & Development
Organisation
201, DRDO Bhawan
New Delhi-110105

FOREWORD

Machineries run by diesel, a component of bio-diesel of fossil fuel creates carbon dioxide emission and pollution and China and India rank among top 5 carbon dioxide emitters in the world. Requirement of diesel by 2011-12 is expected to be 67 million tonnes against 52 million tonnes at present. Contrary to the demand situations, the domestic supply is in position to cater to only about 22% of the total demand and for rest 78% country depends on import of diesel. Thus, energy security has become a national issue and serious attempts need to be made to reduce dependency on imports and seek better alternatives. After the foresight of Rudolf Diesel in 1912 about the vegetable oils to be used as petroleum as an alternate source, many countries started bio-diesel production and consequently Germany is leading country producing 56% bio-diesel of its total demand followed by France (19%) and Italy (17%) from edible oil i.e. rapeseed & sun flower.

Since Country has insufficient production of vegetable oil so the alternate left is looking for non-edible oil produced from *Jatropha curcas*. Keeping in view the demand of 67 million tonnes of diesel by 2011-12 country requires 13.4 million tonnes of bio-diesel at B 20 blending and this will require 13 million ha land already available for Jatropha cultivation.

Jatropha curcas being a hardy plant is widely adopted in different geo-climatic condition of India and can help in energy security besides environmental conservation, rural employment and economy. The book entitled "Bio-diesel: *Jatropha curcas* (A promising source)" is timely written and author's endeavor has covered various aspects of Jatropha cultivation which would prove most useful for the readers.

I take the opportunity to congratulate authors for their untiring efforts made in publication of this book.

W. Selvamurthy
DS, CC R&D (LS &HR)

PREFACE

Unprecedented exploitation of fossil fuel and deforestation has threatened not only environmental hazard but also has opened eyes for exploration of alternate source of bio-diesel production and energy security. India being a country of mega biodiversity has a varieties of tree borne non-edible oil seed trees like *Jatropha curcas, Pongamia pinnata, Azadirachta indica, Simaruba* and *Madhuca indica* etc. Agro-geo-climatic condition of India is most suitable for cultivation of Jatropha because of its varied utility.

Use of tree borne oil as bio-diesel dates back since the discovery of Rudolf Diesel in 1912 & 1913 and Walton (1938). However, massive work on Jatropha came in big way in India in 2003 and in this short span there has been a number of books published on Jatropha and a huge area has been covered by Jatropha plantation. Available books are undoubtedly very useful, however, different aspects on Jatropha cultivation, bio-diesel production and its utilization is still lacking.

Our endeavor, therefore, has been designed to cover all relevant aspects and diverse area in simple style, however, deal all scientific aspects of Jatropha cultivation and bio-diesel production. We do hope it would serve the purpose for which it is intended and will satisfy the readers. We have tried to present the subject matter accurately and clearly based on our experiments, experiences, survey. Several research papers, presentation and personal communication quoted have provided valuable help and support in subject matter. It is hoped readers would not hesitate to help the authors by drawing their attention to any inaccuracies and inconsistencies which may have escaped.We shall be looking forward to receive valuable suggestions, criticisms and comments of readers for the improvement of book.

We are highly grateful to **Hon'ble Sh. Pallam Raju, Raksha Rajya Mantri,** Ministry of Defence, **Dr. W. Selvamurthy,** DS, CC R&D (LS&HR), DRDO, New Delhi for their encouragement to publish a book on Jatropha. We are also grateful to **Dr. Narendra Kumar,** Outstanding Scientist, Director, Directorate of Personnel,

DRDO Bhawan, New Delhi for his encouragement without whom this book has not been possible. Also we are highly indebted to various authors of research papers, articles, literatures, and presentation from where materials were taken for publication of this book.

Our sincere thanks are due to Mr. Pankaj Kumar for designing, data feeding and giving shape to the book.Last but not the least to our readers we request to suggest us improvements and lacunae if any found in our endeavor.

Mohommad Arif
Zakwan Ahmed

CONTENTS

BIO-DIESEL

Jatropha Curcas

(A PROMISING SOURCE)

1. INTRODUCTION

Since the knowledge of *Jatropha curcas* L. as a rich source of bio-diesel, a lot of emphasis has been given and consequently Government and Non Government organization have started planting *Jatropha* and looking forward to produce bio-diesel. Under national Jatropha mission, at present there is a network of about 42 Government Institutes and Agricultural Universities in 21 states under National Oilseed and Vegetable Oil Development Board (NOVOD), Ministry of Agriculture to look into the development of Jatropha cultivar (s) for different agro-climatic zones and their agri-technology. However, the efforts being made on this line is almost unsystematic and unplanned without foresight of the economic and eco-developmental use of bio-diesel and by products. On the other hand identification of users' demand has not been worked out. Since 2003 with the start of massive work on Jatropha, various private organizations have great worry of their seed and saplings raised in large quantity because as on today there is neither any high yielding seed and oil strain nor there is any integration among the organizations working in the field of various aspects like techniques of raising of saplings, seed and bio-diesel production, detoxification of cake, generation of bio gas, and production of bio fertilizer, feed and glycerol etc.

Maximum machineries are run by diesel of fossil fuel origin, which creates Co_2 emission and pollution. China and India rank among top 5 Carbon dioxide emitters in the world are also large importers of petroleum. These countries depends on external energy sources from highly unstable regions which would lead and increase to uncomfortable levels. Energy security thus become a key issue for India. India produces about 22 % oil of its own requirement indigenously and imports 78% of its total requirement. Besides India's import bill is about 100000 crores per year and consumption is about 2% world's oil. Further India and China will account for 43% increases in global oil consumption. Total crude oil import in India accounts Rs. 121500 crore in 2005. Total diesel consumption in India was 53 million MT (78%) which was imported and 11.50 MMT (22%) was indigenously produced.

As per Planning Commission estimates, the requirement of diesel was about 52 million tonnes in 2006-07 and expected to be 67 million tonnes in 2011-2012. Contrary to the demand situation, the domestic supply is in position to cater to only about 22% of the total demand. Therefore, attempt needs to be made to reduce dependence on imports and seek better alternatives. Under bio-fuel mission, India has recently initiated R&D work on production of bio-diesel. A large number of government/ non-government organizations and industries are engaged in cultivation of tree borne oil species like *Jatropha* and *Pongamia*. A few are engaged in extraction of oil and trans-esterification process and testing & evaluation of bio-diesel. However, there is over publicity and euphoria about plantation and yield due to conspicuous lack of any authentic database. As such there are no authenticated high yielding cultivars identified for different agro-climatic zones. However, ICAR, CSIR, DRDO, DBT, DST, TERI and different agricultural Universities have now been stepping up to come out with location/region specific Jatropha cultivars through multi locational trials. Work regarding oil extraction, esterification and evaluation needs to be strengthened to up scaling the existing technology and establishing techno-economic viability. The requirement of diesel by defence forces is huge which becomes an opportunity to work on bio-diesel with availability of resources i.e. land, human and funds.

India is committed to implement Euro-III and Euro-IV diesel fuels norms for which an investment of about Rs. 65,000 crores shall be required mainly for upgrading refinery facilities. The investment of a part of this amount on bio-diesel could ease our achievement of above-mentioned norms as the vigorous evaluation of bio-diesel fuel indicates that the overall smog forming potential, unburnt hydrocarbons, carbon monoxide and sulphur dioxide emissions are reduced significantly by the use of blended/pure bio diesel. The NOx emission has been found to increase marginally for which certain modifications in the engine parameters shall be required.

To overcome this problem DRDO has been taking this challenge in collaboration with CSIR and different Universities and IIT's to produce bio-diesel by planting *Jatropha curcas* on the land available with different DRDO laboratories and Military Farms along with the studies on the performance of different types of engines/vehicles available with Defence Forces at different altitudes so as to develop self reliance and confidence in the field of indigenous production of bio-fuel, to reduce import of fossil fuel and further development of economy and eco sustenance.

2. HISTORICAL ACCOUNT OF BIO-DIESEL

- Trans-esterification of a vegetable oil was conducted as early as **1853**, by scientists **E. Duffy and J. Patrick**, i.e. even before the birth of the diesel engine! This, however, was not for any fuel application.
- **Walton** recommended that "to get the utmost value from vegetable oils as fuel it is academically necessary to split off the triglycerides and to run on the residual fatty acid."
- Belgian patent 422,87, granted on **Aug. 31, 1937,** to G. Chavanne (of the University of Brussels) constitutes the first report on what is today known as bio-diesel. It describes the use of ethyl esters of **palm oil** as diesel fuel. This was probably inspired by the optimum chain length **(cetane)** of fossil diesel. A bus fueled with palm oil ethyl ester served the commercial passenger line between Brussels and Leuven in 1938.
- Research into the use of trans-esterified sunflower oil and refining it to diesel fuel standard was initiated in **South Africa in 1979**.
- An Austrian Company, Gaskoks, obtained the technology from the South African Agricultural Engineers and put up the first pilot plant for bio-diesel in **1987** followed by the erection of the

first industrial bio-diesel plant in **1989**, with a capacity of **30,000 tons** of rape seed per annum.

- 2003 - Biodiesel production touches 1.4 Million Metric Tons in Europe and 80,000 Metric Tons in USA
- 2004 - First demonstration by Daimler Chrysler/CSMCRI of high quality (EN14214) bio-diesel from non-edible vegetable oil from *Jatropha curcas.*

3. BIO-DIESEL

The Bio diesel Association of Canada defines bio diesel as the following :

Bio diesel means the mono alkyl esters of long-chain fatty acids that are derived from animal fats and oils and that meet or exceed the specifications or ASTM D6751and/or EN 14214 or any legal successor thereto. Bio-diesel has proven to be effective as a lubricity additive and for use in automotive engine, home heating system and other equipment designed to use diesel.

Bio-diesel , defined as mono alkyl esters of long chain fatty acids of vegetable oils or animal fats, is a promising alternative fuel for use in compression - ignition engines and is being produced or used commercially in numerous countries around the world. Concept of bio-diesel (bio-fuel) dates back to 1895 when Dr. Rudolf Diesel built the first diesel engine to run on vegetable oil. Bio-diesel is a variety of Ester based oxygenated fuels derived from natural, renewable biological resources such as vegetable oils. Bio-diesel operates in compression ignition engines like petroleum diesel there by requiring no essential engine modifications. Moreover it can maintain payload capacity and range of oil conventional diesel. Unlike fossil diesel, pure bio-diesel is biodegradable, nontoxic and essentially free of sulphur and aromatics.Bio diesel is the generic name of an alternative diesel fuel produced from renewable resources that are converted into fatty acid methyl esters. It is a versatile fuel that can be used as a substitute or additive in a range of diesel fuel applications. Different fats and vegetable oils produce somewhat different bio diesel fuels.

Bio-diesel is a renewable fuel for diesel engines. Bio-diesel defined by ASTM International D 6751, consists of long chain fatty acid alkyl esters and is made from renewable vegetable oils, recycled cooking oils or animal fats. It can be used at full strength, but it is typically blended with petroleum diesel. Technically bio-diesel is vegetable oil methyl or ethyl ester. Bio-diesel molecules are very simple hydrocarbons containing no sulfur, ring molecules or aromatics associated with fossil fuels.

It is an eco-friendly, alternative fuel prepared from renewable resources i.e. vegetable oil (edible or non edible) and animal fat with least emission of green house gases. These natural oil and fat are made up mainly of triglycerides. These triglycerides when treated chemically with lower alcohol in presence of catalyst result in fatty acid esters. These esters have striking similarity to petroleum-derived diesel and are called **"Bio-diesel"**. Earlier this oil was used mainly for candles and soap production. During the Second World War it was used as bio-fuel, substituting for diesel. A few countries use Neem (*Azadirachta indica*), Sal (*Shorea robusta*), Khakan (*Salvctoria oleictos*), Mahua (*Madhuca indica*), Karanj (*Pongamia pinnata*), Kusum (*Schleichera oleosa*), Kokum (*Garcinia indica*) and many other from where bio-diesel is extracted. However, main petro plants belong to family Euphorbiaceae and rich source of bio-diesel is *Jatropha curcas*. Physical characteristic features of bio-diesel are given in Table-1.

Table 1: Physical characteristics of bio-diesel

Sl. No.	Properties	Values
1.	Specific gravity	0.87 to 0.89
2.	Kinetic viscocity @ 40^{0}C	3.7 to 5.8
3.	Cetane number	46 to 70
4.	Higher heating value (Btu/lb)	16928 to 17996
5.	Lower heat value (Btu/lb)	15700 to 16735
6.	Sulphur wt (%)	0.00 to 0.0024
7.	Cloud point ^{0}C	-11 to 16
8.	Pour point ^{0}C	-15 to 13
9.	Iodine number	60 to135

4. THE FIRST BIODIESEL

Walton (1938) recommended that "to get the utmost value from vegetable oils as fuel it is academically necessary to split off the triglycerides and to run on the residual fatty acid. Practical experiments have not yet been carried out with this; the problems are likely to be much more difficult when using free fatty acids than when using the oils straight from the crushing mill. It is obvious that the glycerides have no fuel value and in addition are likely, if anything, to cause an excess of carbon in comparison with gas oil."

Although Walton's statement points in the direction of what is now termed bio-diesel by recommending the elimination of glycerol from the fuel, some remarkable work performed in Belgium and its former colony the Belgian Congo (known after its independence for a long time as Zaire) deserves more recognition than it has received. Indeed, it appears that **Belgian patent 422, 87, granted on Aug. 31, 1937,** to G. Chavanne (1938) (of the University of Brussels), constitutes the first report on what is today known as bio-diesel. It describes the use of ethyl esters of palm oil (although other oils and methyl esters are mentioned) as diesel fuel. These esters were obtained by acid-catalyzed trans-esterification of the oil (base catalysis is now more common). This work has been described in more detail by Knothe (2001) inform Industrial Oils Artificial "gasoline," "kerosene," and "diesel" were obtained in China from tung oil and other oils. Other oils that were used in such an approach included fish oils, linseed oil, castor oil, palm oil, and cottonseed oil.

Of particular interest is a related extensive report published in 1942 on the production and use of palm oil ethyl ester as fuel). That work described what was probably the first test of an urban bus operating on bio-diesel. A bus fueled with palm oil ethyl ester served the commercial passenger line between Brussels and Louvain (Leuven) in the summer of 1938. Performance of the bus operating on that fuel reportedly was satisfactory. It was noted that the viscosity difference between the esters and conventional diesel fuel was considerably less than that between the parent oil and conventional diesel fuel. Also, the article pointed out that the esters are miscible with other fuels. That work also discussed what is probably the first cetane number (CN is a combustion-related diesel fuel quality index) testing of a bio-diesel fuel. CN of palm oil ethyl ester was reported as approximately 83 (relative to a high-quality standard with CN 70.5 and a low-quality standard of CN 18 and diesel fuels with CN of 50 and 57.5). Thus, those results agree with "modern" work reporting relatively high CN for such bio-diesel fuels.

In more recent times, use of methyl esters of sunflower oil to reduce the viscosity of vegetable oil was reported at several technical conferences in 1980 and 1981 and marks the beginning of the rediscovery and eventual commercialization of bio-diesel. A final thought should be given to the term "bio-diesel" itself. A *Chemical Abstracts* search (using the "SciFinder" search engine with "bio-diesel" as the key word) yielded first literature use of the term bio-diesel in a Chinese paper published in 1988. The next paper using that term appeared in 1991 and from then on the use of the term "bio-diesel" in the literature has expanded exponentially.

5. DEFINITION OF BIO-FUEL

- Bio-fuels means liquid or gaseous fuel for transport produced from biomass.
- Bio-fuels are renewable liquids fuels coming from biological raw material and have been proven to be good substitute for oil in the energy sector.
- Biomass means the biodegradable fraction of products, waste and residues from agriculture, forestry and related industries, as well as the biodegradable fraction of industrial and municipal waste.
- Other renewable fuels "means renewable fuels, other than bio-fuels, which originate from renewable energy sources".

6. PRODUCTS CONSIDERED FOR BIO-FUELS

- Ethanol produced from biomass and or the biodegradable fraction of waste, to be used as bio-fuel.
- Methyl ester produced from vegetable or animal oil, of diesel quality, to be used as bio-fuel.
- Other such as biogas, bio-methanol, bio-methyl-ether, bio-ETBE (ethyl-tertio butyl-ether), bio-MTBE (methyl-tertio-butyl-ether), biosynthetic fuels, bio-hydrogen, pure vegetable oil etc.

7. KINDS AND SOURCES OF BIO-FUELS

Kinds and sources of bio-fuels are given in Table 2.

Table 2: Kinds and sources of Bio-fuels

Kinds	Sources
1. Bioethanol	Molasses, Beet, Sorghum, Sugar
2. Biodiesel	Transesterified Vegetable oil
IN FOREIGN COUNTRIES	
Germany	Rapeseed, Sunflower seed
France	Rapeseed
Italy	Sunflower seed, Rapeseed
United States	Soybeans
Denmark	Rapeseed
U.K.	Rapeseed
Brazil	Soyabean and Castor Oil
Malaysia	Palm and Coconut
Spain	Linseed & Olive oils
Greece	Cotton seed
Thailand	Sunflower and Palm oil
Ireland	Frying oil and animal fats
Phillipines	Palm oils
Asia/ Nicaragua, China, Mexico/Africa/ South America	*Jatropha curcas*
IN INDIA	*Jatropha curcas* *Pongamia pinnata*

8. PERSPECTIVES OF VEGETABLE OILS

Vegetable oils used as diesel fuels well before the energy crises of the 1970s and early 1980s generated renewed interest in alternative fuels, however, details on the first uses are often unclear in literature. Available literature reveals that early use of vegetable oils as diesel fuels is that Rudolf Diesel (1912 & 1913), the inventor of the engine that becomes his name tested "his" engine on peanut oil at the 1900 World"s Fair in Paris (Knothe,2001) bears the name of the engine. Diesel continued that "Similar successful experiments have also been made in St. Petersburg with castor oil and animal oil, such as train oil, have been used with excellent results. The fact that oils from vegetable sources can be used may seem insignificant today but such oils may perhaps become in course of time of the same importance as some natural mineral oils and tore products are now. Vegetable oil also was used as emergency fuels and for other purposes during World War II. Brazil prohibited the export of cotton seed oil in order to substitute it for imported diesel (Schwab et al., 1987). Reduced imports of liquid fuel were also reported in Argentina, necessitating the commercial exploitation of vegetable oil (Van den Abeele., 1942). China produced diesel fuel, lubricating oils"gasoline and kerosene" the latter two by cracking process from tung and vegetable oils (Walton, 1938 and Weibe and Nowakowska, 1949). However, the exigencies of the war caused speedy installation of cracking plants based on fragmented data.

Researchers in India prompted by the events of World War II, extended their investigations on ten vegetable oils for development as a domestic fuel. Work on vegetable oils as diesel fuel ceased in India when petroleum based diesel fuel again became available plentiful at low cost. The Japanese battleship Yomato reportedly used edible refined soybean oil as banker fuel. Concerns about the rising use of petroleum fuel and the possibility of resultant fuel shortages in the United States in the years after World war II played a role in inspiring a "duel fuel" project at the Ohio State University (Columbus, Ohio), during which cotton seed oil and corn oil and blends thereof with conventional fuel, were investigated. Again energy security perspective have become a significant during force for the use of vegetable oil-based diesel fuels, although environmental aspects, mainly reduction of exhaust emission, play a role at least as important as energy security. For example in the United States, the clean Air Act Amendments of 1990 and the Energy Policy Act (EPA) of 1992 mandate the use of alternate, or "Clean" fuels in regulated truck and bus fleets. Amendments to the energy policy act enacted into law in 1998 which provide credits for bio-diesel use (also in blends with conventional diesel fuel), are a major reason the use of bio-diesel in the United States is increasing significantly.

Generally, factors such as geography, climate, and economics determine which vegetable oil is of most interest for potential use in bio-diesel fuels. Thus, in the United States, soybean oil is considered as a prime feedstock; in Europe, it is rapeseed (canola) oil; and in tropical countries, it is palm oil. As noted above, different feedstocks were investigated in the "historic" times. These included palm oil, soybean oil, cottonseed oil, castor oil, and somewhat less common oils, such as

Table 3. Worlds Top Bio-Fuel Producers

Country	Amount (Million liters)	Shares of world production (%)	Primary feed stock
Ethanol			
Brazil	15,110	37	Sugarcane
United states	13,390	33	Corn
China	3,650	9	Corn, Cassava, and other grains
India	1,750	4	Sugarcane, Cassava
France	440	2	Sugar beets, wheat
Bio-diesel			
Germany	1,310	56	Rapeseed, sun flower seed
France	440	19	Rapeseed
Italy	400	17	Sunflower seed, rape seed
United States	95	4	Soybeans
Denmark	88	4	Rapeseed
U.K.			Rape seed
Brazil			Soyabean, Jatropha & Jojoba
Malaysia			Palm
Spain			Linseed & Olive
Greece			Cotton seed
Asia/Africa/Nicaragua, Thailand, Zimbabwe, Austria & Nigeria			*Jatropha curcas*
India			*Jatropha curcas & Pongamia pinnata*
China			*Jatropha curcas & Pistacia chinensis*
Mexico			*Jatropha & Jojoba*
Australia			*Jojoba*
Costa Rica			*Jojoba*
Paraguay			*Jojoba*
Argentina,Ezypt, Israel & Peru (Commercial Plantation)			*Jojoba*

Several countries in the world have active bio-diesel programme with legislative support and have national polices on bio-diesel development. Soya is used for bio-diesel production in USA, Rape seed in France, USA and Germany, Sunflower in Italy and Southern France, Castor in Brazil, Coconut in Malaysia, Palm in Thailand and Philippines, Linseed and Olive in Spain, Cotton in Greece, Jatropha in Nigeria. Whereas, countries like Ireland, Austria, the Czech Republic, Denmark, Italy and Sweden have frying oil and animal fat for conversion to bio-diesel.

Bio-diesel production has registered a substantial leap during last 10 years, in **EU (European Union)** countries viz., Germany, France,

Italy and Denmark. During this period, world wide production has increased to over 3.5 MMT. European union enforced several legislative action to double the renewable energy sources. The most important of these is to replace diesel and petrol up **to 5.75% for transportation by 2010.** EU has taken a lead in the production of bio-diesel across the world . There are fuel dispensing units for B5, B10, B20 & even B100 at retail outlets. Some EU countries have regulations for mandatory use of bio-diesel. Pricing of bio-diesel is comparable with petro- diesel and is widely accepted by main OEMs.

Bio-diesel to be marketed internationally must meet the high standard set by the ASTM D6751 (US/Canada). The European specification EN14214 (Europe/Israel) is similar but slightly more stringent than ASTM D 6751 in a few areas. Therefore, the ASTM D6751 is the minimum acceptable specification for B100 bio-diesel stock. Bio-diesel specification of India is IS 15607, Argentina-IRAM 6515, China GBIT-20828 and Indonesia-SNI 04-7182. Production scenario in few bio-diesel leading countries is presented below.

Germany

The rape seed bio-diesel is produced in Germany. The world's largest producer of bio-diesel has about 50% share in world production. Petro-diesel in Germany blended with 2-5% bio-diesel will be soon followed by whole of Europe. It has evaluated toxic and non-toxic cultivars, studied chemical composition of seed, digestibility, protein degradability and toxicity factors at University of Hohenheim (Makkar et. al., 1996) · Studied ethnopharmacology at German Institute of Medical Mission (Muller and Mechler, 2005) and evaluated performance of bio-diesel in engines (Pak and Allexi., 1994). University of Tubingen is applying NMR spectroscopy for sound accuracy of reproducibility and identification of oil (Akinttayo and Bayer, 2002). Preliminary work on detoxification of Jatropha cake has been done at the University of Hohnenheim applying heat, sodium hydroxide and sodium hypochlorite method (Aregheore et al., 2003), However, it has not achieved commercial maturity. Germany has more than 1500 bio-diesel filling stations.

France

France is the world's second largest producer of bio diesel. Its conventional diesel contains between 2-5% bio diesel and that will soon apply to the whole of Europe. Being the second largest producer with 19% share accorded national priority for bio-diesel production with target to blend all fuels with 5.75% bio-diesel by the end of 2010. The country has sown 1 million ha with rape seed against a target of 0.70 million ha (www.planetark.com/dailynewsstor.cfm/newsid/30888/story.htm). Sunflower based bio-diesel has made good success in France.

United States

Soya based bio-diesel is produced in USA. US National Energy Policy is to increase its energy supply using more diverse mix of domestic resources to reduce dependence on imported oil. The new energy bill mandate is to use over 10% of renewable fuel by increasing production from its present share of 4% in world bio-diesel production. Its bio-diesel programme is mainly based on soybean, canola and waste cooking oil from restaurants. Bio-diesel is receiving substantial political support in US from the Congress and the Senate in creation and improvement of general taxation and administrative conditions. The Energy Ministry of US has recognized bio-diesel as an alternative fuel for vehicle fleets as specified by the Energy Policy Act and Environment Protection Agency.

The most advanced US military is also facing one of the greatest challenges in energy security and has initiated efforts to have alternate source. Defence Advanced Research Projects Agency (DARPA), Pentagon is looking for a new domestic bio-jet fuel derived from agriculture or aquaculture crops as an alternative to military jet fuel (JP-8) (reentech.htm). The Task has been assigned to academia, the University of North Dakota. JP-8, a petroleum based fuel, is used to power vehicles in the U.S. military such as the Boeing B-52 bomber, the Abrams A1 Battle Tank, the Apache Helicopter, and many others. The U.S. Army, Navy, Air Force and Marines all use bio-diesel blend, B20 at its different bases and stations throughout the country (www.biodiesel.org). Bio-diesel had similar horse power, torque and BTU content compared to petroleum diesel. It offers excellent lubricity and higher cetane than diesel fuel. Bio-diesel testing revealed that older equipments took a filter change, but newer equipment needed nothing. The U.S. Department of Navy recently announced a new policy that will lead to greater use of the domestically produced fuel and increase U.S. energy security by reducing dependence on foreign sources of oil. Principal Deputy Assistant Secretary (Installations and Environment) Wayne Army had issued a memorandum that establishes a policy that all U.S. Navy and Marine non-tactical diesel vehicles shall operate on a blend of 20% bio-diesel fuel (B20).

Bio-diesel blend performance has been tested by several important measures and found to give better results than petroleum diesel. But its relatively high production cost and limited availability of raw materials used in its production continue to limit its commercial application. Storage in cold climate and testing studies revealed that the bio-diesel should be kept at least 10 ^{0}F above its cloud point to successfully blend with diesel. National Renewable Energy Laboratory (NREL) has sponsored research to find bio-diesel formulations that do not increase nitrogen oxide emissions by adding cetane enhancers- di-tert-butyl peroxide at 1 percent or 2-ethylhexyl nitrate at 0.5 percent. Use of kerosene to reduce nitrogen oxide emissions from blends containing 20 percent bio-diesel is under investigation (www.eia.doe.gov/oiaf/analysispaper/biodiesel/index.html).

A storage study completed over a 24-month period at University of Idaho found that bio-diesel tends to store about as well as diesel fuel. This study found that engine power decreased about 2% and viscosity, density, peroxide and acid value increased for bio-diesel. Usually it is recommended not to store bio-diesel longer than 6 months or at the most, a year. This recommendation is similar to diesel fuel storage periods. NREL is performing nationwide survey of bio-diesel and bio-diesel blend quality in order to reveal engine manufacturers and fuel consumers about the engine and fuel system component durability. NREL is assessing impact on reliability by doing fuel pump and fuel injector wear tests along with materials compatibility tests. Use of suitable anti-gel or pour point depressant have helped in satisfactory engine operation when the ambient temperature is sub zero. There is a range of anti-gel additives made for use with petro-diesel, but these do not work very well with bio-diesel. The additives specially formulated for use with bio-diesel include Wintron XC30 (Bio-fuel System Ltd) (www.biofuelsystems.com), Arctic Express Bio-diesel Anti-gel (Power Service Products) (www.powerservide.com) and Lubrizo (Pour point depressant for Vegetable oil and bio-diesel) (www.lubrizol.com).

US Defence Energy Support Center (DESC) is the watchdog agency that monitors the quality of military fuels. DESC has strong concerns about the quality control of the bio-diesel blends that are currently available for sale. If all US military vehicles, aircraft, and vessels must one day move to bio-diesel, the sheer volume will mean that the forces will have to buy from civilian suppliers.

Denmark

Denmark is actively engaged in R&D on energy crops and production of bio-fuels. Danish consumption of bio-fuels for transport is very low and limited to experiments at local level. Hence, the major portion of bio-diesel produced is exported. Its share in world bio-diesel production is about 4% which is based mainly on rapeseed.

Brazil

Brazil responded timely to increase oil price in international market and executed "Proalcool" programme successfully during seventies. It was the first and most successful large scale bio-fuel programme in the world and addressed several objectives, making Brazil a world leader in bio-ethanol production and use. Attention is also being paid for cultivation of Jatropha, its tissue culture (Jesus et al., 2003) and agro- technology (Saturnino et al., 2005). Jatropha has been identified as a potential traditional oil crop among soybean, cotton, sunflower, *Ricinus communis* and oil palm (Teixeira, 2005).

China

The bio-diesel development in China is gradually progressing. Major oil bearing plants found in China are *Jatropha curcas, Pistacia chinensis, Cornus wilsoniana* and *Xanthocerus sorbifolia. Jatropha curcas* has been extensively exploited for medicinal use and toxic properties (Lin, Juan, et al., 2004), standardization of tissue culture protocol (Wei et al., 2004) and identification of curcin in calli generated from Jatropha explants (Rong et al., 2005) . Super critical method has been identified as the best for oil extraction over microwave and ultrasonic methods due to its comparatively higher yield with best quality of oil which does not need refining (Zeng et al., 2005).

Thailand

Jatropha curcas is the identified energy plant for production of bio-diesel in Thailand. Studies on propagation techniques (Stienswat et al., 1986) and plant geometry (Thitithanavanich,1984 and Ratree, 2004) for optimum yield have been conducted. Trans-esterification process (Stienswat et al., 1986 a & b) using 0.5% NaCl and 1.0% $NaOCH3$ has been standardized. Fatty acid composition and properties of Jatropha oil and methyl ester have been established (Chatakanonda, 2005).

Austria

Graz University of Technology has been working on *Jatropha curcas* since nineties (Gubitz, et al., 1999) and successfully used cell wall degrading enzymes like cellulolytic and hemicellulolytic with alkaline protease for higher oil recovery through aquous extraction (Winkler, et al., 1997) besides development of trans-esterification process. A semi-continuous Upflow Anaerobic Sludge Blanket (UASB) process has been standardized for bio-gas production from pressed cake (Staubmann, et al., 1997).

Zimbabwe

The Agricultural Research Trust, (ART) Zimbabwe has developed non-toxic varieties of *Jatropha curcas*. The seed cake of this variety is suitable as animal feed without detoxification. However, its authenticity is under evaluation. Midlands State University had tried to bring phorbol esters and anti-nutritional factors to safer level through various methods but could not succeed to detoxify phorbol esters comparable to non toxic variety i.e. 0.11 mg/g (Chivandi, et al., 2004).

Nigeria

Jatropha curcas is used as a potential source of bio-fuel (Foidi, et al., 1996). Extraction, characterization (Ajiwe, et al., 1996) and analysis of seed for fatty acids, lipids and sterols has been carried out. Physico-chemical analysis of oil has been studied to find out its application in alkyd resin and soap manufacture (Abigor and Uadia, 2001).

Italy

Italy falls among top five bio-diesel producers in world with about 17% share which is based on soybean, sunflower and rape seed. Use of bio-diesel in boilers and internal combustion engines has been sucessfully done (Carraretto, et al., 2004). B100 and blends were used in ICE with varying results. Initial studies on combustion dynamics have also been carried out with promising results in reducing urban pollution.

South Africa

The cultivation and management of Jatropha is poorly documented in South Africa and there is little field experience available. Currently, growers are unable to achieve the optimum economic benefits from the plant.

momentum, however, there are certain technological constraints which need attention in this field.

As per statistics on land availability, the total geographical area of India is 329 million hectares. Out of this, area under cultivation is about 173 million hectares. It is generally estimated that large area of the geographical area of the country is either waste or degraded. This area can be used for cultivation of bio- fuel plants. National Policy is being formulated to ensure sustainable production, conversion and application of bio-fuel to reduce the increasing and worrysome on us of crude oil imports (Anonymous, 2006). The National Bio-fuel Development Board (NBDB) which will work under Prime Minister, comprises of Ministry of Rural Development, Ministry of Agriculture, Ministry of Non Conventional Energy Sources, Ministry of Petroleum and Natural Gas, Department of Biotechnology, Science and Technology, and the Planning Commission (Anonymous, 2006). According to the report, the short term target is to replace 5% petro-diesel by 2012, 10% by 2017 and 20% by 2030. Few states like Uttaranchal, Chhatisgarh, Andhra Pradesh have constituted bio-fuel boards / Bio-fuel Development Authorities at state level.

Ministry of Petroleum and Natural Gas commission has also laid down a bio-diesel purchase policy (www.pcra.org/r&d/purchase.htm) which recognizes the vital role that Panchayati Raj Institutions can play in promotion of bio-diesel in consultation with NOVOD Board, Department of Biotechnology and National Botanical Research Institute which are working on a network programme for quality plantation. The Petroleum Conservation Research Association (PCRA) has established a National Bio-fuel Center at its HQ in New Delhi to provide information on bio-fuel development.

12. TECHNOLOGIES CONSTRAINTS FOR BIO-DIESEL

There is still big gap in technologies available/not available on Jatropha cultivation. This has to be taken care of to meet the National policy on bio-fuel so as to achieve the target and make the venture, viable, useful, economic, employment opportunity and eco-friendly.

AVAILABLE	NOT AVAILABLE
• Oil extraction	• Non-toxic cultivar
• Transesterification	• Process for by-products
• Chemical analysis	• Detoxification of cake
• Testing in engines	• Extension of shelf life

13. BIO-DIESEL DEVELOPMENT IN INDIA

It would be possible to substitute petro-diesel with bio-diesel by 20% without any alterations/modifications in existing CI engines. Thus, to have 20% blend of 67 million tonnes petro-diesel, 13.40 million tonnes of bio-diesel will be needed which in turn will require about **13 million ha of land**. This can yield a revenue of approximately 20,000 crore rupees per annum besides employment opportunities to over 12 million people in cultivation and extraction/ Trans-esterification operations. The *Jatropha curcas* is available in almost all the states but in scattered manner and that too with traditional plantation with low productivity and less oil content in the seed. Countrywide efforts have been made to identify superior trees having more yield and high oil content. The seed material of promising cultivars has been collected and is being provided to various Agricultural Universities and its Institutions for further evaluation and developing model plantation. Such plantation has been undertaken in more than 10,000 ha by 37 different Universities, Institutes, IIT's, SAU's and NGO's in 24 states. However, as per estimates of NOVOD about 31.17-lakh ha plantation would be completed by the end of 2008-09 to substitute about 5% diesel requirement of entire country (Anonymous, 2006). The country has an estimated potential of more than 50 lakh tons of tree borne oil seeds with an annual potential of 2 million tonnes, however, only 8.10 lakh tonnes are being collected. At present 1.5 to 2.0 lakh tonnes of oil is being extracted out of 10 lakh tonnes of oil seed.

Under bio-fuel mission, India has recently initiated R&D work on production of bio-diesel. A large number of government/non-government organizations and industries are engaged in cultivation of tree borne oil species like Jatropha and *Pongamia*. A few are engaged in extraction of oil and trans-esterification process and testing & evaluation of bio-diesel. However, as such there are no authenticated high yielding cultivars identified for different agro-climatic zones. Work regarding oil extraction, esterification and evaluation needs to be strengthened to up scaling the existing technology and establishing techno-economic viability.

14. EXPECTED DIESEL / BIO-DIESEL DEMAND IN INDIA

In India, about 52 million MT of diesel fuels was consumed in 2006-2007 and this is likely to exceed 67 million MTs in 2011- 12. The estimated diesel fuel demand for the next twenty years along with the requirements of bio diesel for B5, B10 and B20 blend levels is given in Table 5. For 20% blending by 2030, 38 million ha of waste land is required if yield of Jatropha is estimated 5 tonnes /ha.

Table 5. Diesel and Biodiesel demand in India

S. No	Year	Diesel Demand million MTs	Bio-diesel Demand (Million MTs)		
			B_5	B_{10}	B_{20}
1.	2006-07	52.33	2.62	5.24	10.48
2.	2007-08	55.26	2.76	5.52	11.04
3.	2008-09	58.35	2.92	5.84	11.68
4.	2009-10	61.62	3.08	6.16	12.32
5.	2010-11	65.07	3.25	6.50	13.00
6.	2011-12	66.90	3.35	6.69	13.08
7.	2020-21	111.92	5.60	11.20	22.38
8.	2030-31	202.84	10.14	20.28	40.56

To produce bio-diesel at 5%, 10% & 20% blending there is requirement of 2.79, 5.58 & 11.19 mha area (Table 6.a).

Table 6.a: Area coverage and blending requirements

Year	Diesel Demand MMT	Area required for blending (mha)		
		5%	10%	20%
2005-06	49-56	2.07	4.14	8.28
2006-07	52.33	2.19	4.38	8.76
2007-08	55.26	2.48	4.68	9.37
2008-09	58.35	2.77	4.98	9.98
2009-10	61.62	3.06	5.28	10.59
2010-11	66.90	3.35	5.58	11.19

Source: R. Mandal (2005) "Energy- Alternate solutions for India's needs: bio-diesel" Advisor for the Planning Commission, Govt. of India, 2005.

15. BIO-DIESEL SPECIFICATION

Most of the advanced countries have finalized their bio diesel specification. In India, Bureau of Indian Standards (BIS IS-1460-2000) is in the advanced stage of finalization of standard for B100 Bio diesel fuels. The same specifications shall be applicable for bio diesel for blending. Similarly, OEM approvals for blended and B100 bio diesel are in the process through pilot level trials under way at different places.

Bio diesel has proven to be effective as a lubricity additive and for use in automotive engines, home heating systems and other equipment designed to use diesel fuel. Any product marketed as bio diesel must meet the high standard set by the American Society for Testing and Materials (ASTM) D6751. The European specification EN14214 is very similar to the ASTM D6751, and is in fact slightly more stringent in a few areas. Therefore, the ASTM D6751 is the minimum acceptable specification that is acceptable as B100 bio diesel blend stock. In its neat form as B100 (100% bio diesel), bio diesel offers significant environmental benefits. In absence of a national standard, the US, military has developed its own specifications for B20 bio-diesel blends A-A-52557 according to which bio-diesel standards were set conforming to ASTM D 6751 and a balance diesel fuel conforming to ASTM D 975 or to the military specification. The standard further specifies a number of properties and test methods for B20 blend. While military standard does not include a fuel stability specification, it advises against using bio-diesel blends that have been stored for longer than six months from the date of manufacture.

ASTM International is one of the largest voluntary standards development organizations in the world-a trusted source for technical standards for materials, products, systems, and services. Known for their high technical quality and market relevancy, ASTM International standards have an important role in the information infrastructure that guides design, manufacturing and trade in the global economy. ASTM International, originally known as the **American Society for Testing and Materials (ASTM)**, was formed over a century ago, when a forward-thinking group of engineers and scientists got together to address frequent rail breaks in the burgeoning railroad industry. Their work led to standardization on the steel used in rail construction, ultimately improving railroad safety for the public. As the century progressed and new industrial, governmental and environmental developments created new standardization requirements, ASTM answered the call with consensus standards that have made products and services safer, better and more cost-effective. The proud tradition and forward vision that started in 1898 is still the hallmark of ASTM International.

Today, ASTM continues to play a leadership role in addressing the standardization needs of the global marketplace. Known for its best in class practices for standards development and delivery, ASTM is at the forefront in the use of innovative technology to help its members do standards development work, while also increasing the accessibility of ASTM International standards to the world. ASTM continues to be the standards forum of choice of a diverse range of industries that come together under the ASTM umbrella to solve standardization challenges. In recent years, stakeholders involved in issues ranging from safety in recreational aviation, to fiber optic cable installations in underground utilities, to homeland security, have come together under ASTM to set consensus standards for their industries. Standards developed at ASTM are the work of over 30,000 ASTM members. These technical experts represent producers, users, consumers, government and academia from over 120 countries. Participation in ASTM International is open to all with a material interest, anywhere in the world. (http://www.astm.org/about/about.astm.html).

Bureau of Indian standards (BIS) also adopted specifications for bio-diesel for use in India is ISI 5607(2005). The greatest virtue of bio diesel is that it contains virtually no sulphur. Furthermore, according to a report issued by the EPA in October 2002, burning B100 reduces the emissions of particulate matters (PM) and carbon mono oxide (CO) by almost 50% and unburnt hydrocarbons by almost 70%. There is, however, a slight increase in Nox emissions, but blending reduces Nox emissions to a negligible amount. Research is currently being conducted to lower these Nox emissions. According to the **National Bio-diesel Board (NBB)**, bio diesel is the only alternative fuel to have fully completed the health effects testing requirement of the U.S.1990 Federal *Clean Air Act* Amendments, which required a four-year, $2 million health effects testing programme. Bio diesel is a legally registered fuel and fuel additive with the Energy Policy Act (EPA) and is a legal fuel for U.S. commerce. Bio-diesel specification as per Indian standard 15607 in comparison to Jatropha bio-diesel is given in Table 6b.

Table 6.b: Comparative specification of Bio-diesel and Jatropha Bio-diesel

S. No.	Property (units)	IS-15607 units	Jatropha Bio-diesel	Significance
1.	Flash Point (^{0}C)	Min 120	137	Temperature at which fuel produce ignitable vapor-air mixture on exposure to a flame.
2.	Phosphorus (mg/kg)	Max 10	<10	It damages catalytic converters used in emissions control system
3.	Water content (mg/kg)	Max 5000	210	It causes corrosion of tanks and other metallic parts/equipments.
4.	Density at 15^{0}C (kg/m^3)	500-900	880	It is a measurement of a mass per unit volume
5.	Carbon residue (Ramsbottom) (% mass)			It is a measurement of carbonaceous material left in fuel after all volatile components are vaporized in absence of air
6.	Sulfated Ash (% mass)	Max 0.02	0.01	It check the presence of residual catalyst and other impurities.
7.	Viscosity at 40^{0}C (ast)	2.5-2.6	4.1	Measurement of fluid's resistance to flow.
8.	Sulfur (mg/kg)	Max 50	53	Measurement of ignition quality of fuel
9.	Cetane member	Min 51	53	Measurement of ignition quality of fuel
10.	Copper corrosion (3 hrs at 50^{0}C)	Max -1	1	Measurement of possible difficulties with Cu and brass/bronze parts of fuel system
11.	Acid value (mg/KOH/g)	Max 0.50	0.12	Measurement of total acidity of fuel and indicates its tendency to corrode metals. Its high value in bio-diesel increases fueling system deposits.
12.	Free Glycerine (%mass)	Max 0.02	0.018	Checks byproduct glycerin in bio-diesel and causes injector deposits and logs fueling system.
13.	Total Glycerine (% mass)	Max 0.25	0.13	It induces free glycerin's and glyceride portion of any un reacted/partially reacted oil/fat. It also causes injector deposits and adversely affects-cold weather operation.

16. ADVANTAGES OF BIO-DIESEL

There are direct/indirect advantages of bio-diesel given below:

- Ultra low sulphur.
- Low emission of carbon compound.
- Better lubrication and odourless.
- High cetane value.
- No modification of engine upto B20
- Non toxic, four time faster than conventional bio-diesel and biodegradable
- Renewable source of energy
- It is safer to store because of its higher flash point.
- Conservation of natural fossil fuel resources.
- Reduction in pollution by 25%.

country is 32.87 m ha and @ 40% area as waste/marginal land it is 13.2 million ha of land available for Jatropha plantation from where 13.93 m tonnes of oil can be produced calculating minimum 30 q seed yield/ha containing 35 % oil yield. However, with good agricultural practices, yield can be raised to 75 q/ha and the prospects of the oil will be 2.5 times higher. At present total demand of diesel is 52.33 million metric tonnes approx. In case all marginal land is utilized India can harvest 13.2x2.5=33.0 million metric tonnes of Bio-diesel which require 13 million ha land in Jatropha plantation (Table-7). So far total bio-diesel production in India is 30000 MT.

In India Madhya Pradesh, Rajasthan, Maharastra, Andra Pradesh, Karnataka, Uttar Pradesh & Gujarat are the main states where huge waste land is available (Fig. 4) and can be best utilized for Jatropha plantation towards growing fuel energy through green plants.

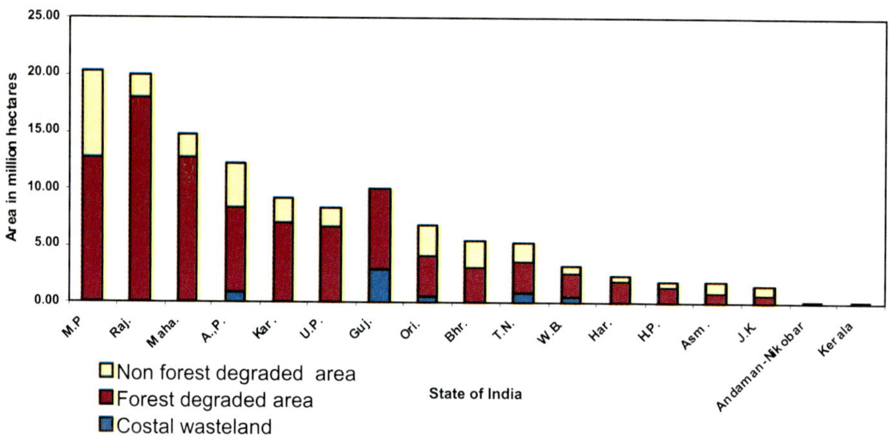

Figure 4: State wise extent of wasteland (in million hectares)
(*Source:* CSMCRI presentation)

Table 7: Waste land availability.

S. No.	Land type	Area (in million acres)
1.	Under stocked forest	7.41m acres
2.	Agro-forestry hedges	12.35m acres
3.	Fallow lands	10.86 m acres
4.	Public Land	2.47 m acres
5.	Railway / Road trades	2.47m acres
	Total	**33.09 m acres** **= 13.4 m ha**

21. PETRO-PLANTS: SOURCE OF BIO-DIESEL

Among all the alternatives, tree borne non-edible oil seed holds a promising prospect in India for production of Bio-fuel as a substitute of fossil-fuel in meeting the future requirements. Plantation of tree borne oilseed species helps to develop eco-system and reduces carbon load from environment by utilizing (i) carbon dioxide during photosynthesis and (ii) emitting remarkably less green house gases than petro-diesel. This helps in maintaining the ozone in the atmosphere. However, plant species grow in unutilized land is to be selected to save the farm land for food security.

Plants are one of the best treasures and source of Bio-fuel and at present a number of countries are using oil of soybean, sesamum and mustard oil as a Bio-diesel. There are a number of trees viz., Sal (*Shorea robusta*), Neem (*Azadirachta indica*), Khakan (*Salvctoria oleictos*), Mahua (*Madhuca indica*), Karanj (*Pongamia pinnata*),Rubber plant, Kusum (*Schleichera oleosa*), Jogoba (*Simmondsia chinensi,)*, Kokeem (*Garcinia indica*) and many other from where Bio-diesel is extracted out. However, mainly petro plants belong to family euphorbiacea and rich source of Bio-diesel is Jatropha. Known species are *Jatropha curcas, J. gossypifolia, J. glandiflora, J. multifilla, J. gigantica, J. hectata, J. tangorensis, J. podagrica and J. integerima.*

Botanical Survey of India has identified over 300 species with non-edible oil rich seeds. As India has chronic shortage of edible oils and imports, non -edible oils are recommended for using in transport and other sectors. Species with bio-diesel potential include Pongamia, Jatropha, Neem, Mahua and Rubber plant. About 450 species of oil yielding plants have been identified in various parts of India and *Jatropha* has been selected for focused development in the country due to less gestation period besides its varied uses. A list of trees bearing seed is given in Table 8.

Table 8: Trees bearing oil (TBO) seeds non edible.

NAME OF PLANTS	OIL PERCENT (%)
Forest Trees -Non edible	
Pongamia pinnata (Karanja)	30-40
Pongamia glabra	30-40
Azadirachta indica (Neem)	20-50
Simmondsia chinensis (Jojoba)	50

contd....

Vateria sp (Dhupa)	20-27
Calophyllum inophyllum (Undi)	50-73
Mesua ferrea (Nahor)	40
Aleurites sp (Tung)	50-60
Schleichera oleosa (Kusum)	34
Ceiba pentandra (Kapok)	20-25
Madhuca latifolia	34-37
Madhuca longifolia	34-37
Madhuca indica (Mahua)	
Madhuca butyracea(Chiura)	
Salvactoria oleidos (Khankan)	
Salvactoria sercia (Khankan	
Shorea robusta (Sal)	
Garcinia indica (Kokum)	
Simaruba glauca (Paradise tree)	20-30
Pistacia chinensis	
Ceernus wisoniana	
Xanthocerus sorbifolia	
Salvadora olecides (Pilu)	
Hevea brasiliensis (Rubber seed)	
Diploknema butracea	
Mallotus phillipines (Kamla)	
Emblica officinalis (Amla)	
Carthamus tinctoris (Kusum)	
Terminalia bellerica (Beheda)	
Simecarpus anacradium (Bhilawa)	
Guizetia abyssinoca(Niger)	
Psorolea corylifolia (Bawachi)	
Carum roxburghianum (Ajmandia)	
Tamarind indicus	
Petro Plants-Edible & Non edible oil	
Glycine max (Soybean)	
Helianthus (Sunflower)	
Arachis hypogea (Ground nut)	
Coccus nucifera (Coco nut)	
Brassica napus (Rapeseed :Canola)	
Brassica compestris (Mustard)	
Petro Plants-Lower Trees and Bushy Plants	
Euphorbia abyssinium	

Euphorbia tritucalli	
Euphorbia resinifera	
Euphorbia lathyrus	
Jatropha curcas (Ratanjyot)	30-40
Jatropha multifida	
Jatropha hastata	
Jatropha gignatica	
Jatropha procera	
Jatropha glandulifera	
Jatropha gossypifolia	
Jatropha tangorensis	
Jatropha podagrica	
Jatropha integerima.	
Jatropha sinera	
Jatropha villosa	
Jatropha maheswarii	
Jatropha diyoka	
Jatropha microfayala	
Jatropha nama	
Jatropha vramnadensis	
Jatropha hetrophylla/ heynei	
Jatropha acrocurcas	
Jatropha mulindnifera	
Petro Plants-Algae	
Botryococcus sps.	
Botryococcus braunii	
Chlorella pyrenoidosa	

Among the trees Jatropha holds promising future due to its wider adaptability, high oil content and multipurpose utility. Plant can be cultivated in variety of soil and thus waste land can be brought under plough for bio-diesel production towards the national energy security, economy, eco-development and consequently self reliance and confidence. *Jatropha curcas* seeds contains approximately 35 to 40 percent oil which can be used as diesel fuel.

The word petro plants originated from the work of Calvin in University of California in 1979, when he reported the collection and use of photo synthetically produced hydrocarbons from the plant. There are some species of plants of certain families, which accumulate the photosynthetic products (Hydrocarbons) of high molecular weight. Calvin suggested it as a substitute of conventional petroleum. Calvin (1979) screened most of the plants of Euphorbiacae specially Euphorbia (2000 species) which reduce CO_2 beyond the hydrocarbons. These plants produce the petrochemicals in the form of milky latex or as seed reserve. The latex is rich in hydrocrakable hydrocarbons, normally called as biocrude, which yields about 70.6% energy. Out of which 22% is kerosene and 44.6% is gasoline (Bio-fuel). About 400 plant species, belonging to different families are known which grow in the different part of country in which Euphorbiaceae has been extensively screened and has shown the fruitful results. Moreover the members of Euphorbiaceae posses higher amount of hydrocarbons than related class of families.

The first gasoline refinery came in to existence in Italy to tap vegetative gasoline in which *Euphorbia lathyrus* an annual herb and *E. tirucalli* a perennial one were used to produce gasoline. *E. lathyrus* can produce 20 tonne, dry matter/ha/year. Chemical analysis of this plant in organic solvents revealed that heptanes extract and ether soluble fraction constitutes about 8% terpenoid extract. By using zeolite catalyst it could be converted into high grade transportation fuel. Of the 85% converted material about 10% was in the form of natural gas and 75% as gasoline fraction (Nemathy et al, 1940). Calvin and co-workers estimated that 10 tonnes of biomass could yield 5.3 barrels of crude extract convertible to gasoline.

21.1 Micro Algae

Micro algae are remarkably efficient biological factories capable of taking a waste form of carbon (CO_2) and converting it into a high density liquid form of energy. The four most abundant classes of micro algae are diatoms (Bacillariophyceae), green algae (Chlorophyceae), blue-green algae (Cyanophyceae), and golden algae (Chrysophyceae). Few species of micro algae have been isolated and identified from N.E. region for potential applications as bio-fertilizers, but not for bio-diesel. They are namely, *Anabaena spiroides, Calothrix marchica, Nostoc linckia, Nostoc punctiforme, Nostoc masconum, Autosira fertilissima, Mastigocladus laminosous, Microchete acqualis, Rivularia hamsgorgii* etc. Biocrude from dead refuse (HCs gasoline) has been reported from fresh water algae viz., *Chlorrella pyrenodids*. Biocurde of good quality but biomass availability is a problem because much R&D work has not yet been done for biomass production. Further fresh water available is going to be a limiting factor.

The cultivation of algae to harvest oil for bio-diesel has not been undertaken in our country since algaculture is an entirely new field to this venture. Enhanced Bio-fuels and Technologies Ltd (EBT), a UK based industry has initiated alga culture work at Coimbatore. Defence Research Laboratory (DRL), Tezpur has collected a few micro algae strains and is progressing towards standardization of cultivation protocol.

Bio-diesel fuel and fossil fuel essentially originated from the photosynthesis products of green plants. Any of a group of organic compound that includes sugars, starches, celluloses and lipids is produced by photosynthetic plants. Heterotrophic organism, including most microorganism and all animals, depend on plants photosynthesis to supply their food and energy. All lipids formation of live-forms is also converted from organic compound of plant photosynthesis products. In principle, it is possible that any kind of fatty acid obtained from plant, animal and microbe are sources of bio-diesel. So the plants or microbe possessing of high lipids content are potentially the best source of bio-diesel.

Benefits of micro algae as material for bio-diesel micro algae are one of the oldest living organisms. They are unicellular species which exist individually, or in chains or groups. Depending on the species, their sizes can range from a few micrometers (μm) to a few hundreds of micrometers. As a group of lower plant, micro algae live annually in oceans, rivers and lakes. Moreover, they have been long known to be essential components of coral reefs. Micro algae comprise a vast group of photosynthetic, heterotrophic organisms which have an extraordinary potential for cultivation as energy crops. Cyanobacteria of micro algae, so-called blue algae or blue greens, are gram-negative photoautotrophic prokaryotes which can be cultivated without nutrition of carbon and nitrogen. The advantages of such cyano bacteria make that they are generally suitable for a low investment, very cost-effective and easy to manage. Micro algae can be cultivated under aqueous conditions ranging from freshwater to situations of extreme salinity and are able to produce a wide range of commercially interesting byproducts,, such as fats, oils, sugars and functional bioactive compounds. Large-scale algae production can be achieved in open raceway ponds or photo bioreactor systems. Micro algae cell growth time of duplicating is in 24 h and even in 3.5 h during the logistic phase of cell multiplication. Micro algae are remarkably efficient biological factories capable of taking a waste (zero-energy) form of carbon and converting it into a high density liquid form of energy (natural oil). Micro algae-based biodiesel has emerged as a viable resource. Micro algae may contain significant quantities of lipids (fats and oil) with compositions similar to those of vegetable oils.

The photosynthesis productivities of hydrocarbons and lipids in some eukaryotic diatom, green algae and brown algae isolated from ocean and lakes reach as high as 50 g/m2/d. The lipids of these algal species are also rich in essential fatty acids, such as C18 and C16 acids and their C20 derivatives. While it is common to find levels of 20%–80% lipids on a dry basis, on occasion the quantities of lipids found in micro algae can be extraordinarily high. For example, in one particular species, *Botryococcus*, concentration of hydrocarbons in the dry matter may exceed 80%, under certain conditions. Meanwhile, the levels of 15%-30% lipids on a dry basis were found in vegetable oils. Chinese mainland coastline stretches for some 18 000 km with a large swamp wetland and marshes, suitable for large scale cyclic cultivation of oil-bearing micro algae. Because of the advantages of its enormous groups, gigantic biomass and very cost-effective management, micro algae and its based bio-diesel are becoming a remarkable source of new bio-fuel and one of developmental trend in clean energy in the future.

21.2 Pongamia pinnata

Pongamia pinnata is one of the few nitrogen fixing trees to produce seeds containing about 30 percent oil. It is often planted as an ornamental and shady tree but now-a-days, it is considered as alternative source for Bio-diesel. This species is commonly called pongam or karanja. Unlike Jatropha, not much attention has been paid to *Pongamia* which is also a promising Tree Borne Oil Seed (TBOs) species for production of bio-diesel. SuTRA Bangalore, TNAU, Coimbatore and Mint Bio-fuels, Pune have established organized plantations during the last 2-3 years. (http://www.nabard.org/roles/ms/fw/pongamia.htm). Among the many species, which can yield oil as a source of energy in the form of bio-diesel, *Pongamia pinnata* has been found to be one of the most suitable species due to its various favourable attributes like its hardy nature, high oil recovery and quality of oil, etc. It can be planted on degraded lands, farmer's field boundaries, Wastelands/fallow lands. Indigenous production of *Pongamia* oil will save foreign exchange worth of several million dollars and also generate employment opportunities in rural areas.

The origin of Karanj being Indian, makes it ideal for cultivation across the country. It is a tree which remains green during summer, adding natural beauty, coolness and provide shelter for bird fauna. It can be grown in water logged areas where Jatropha does not survive. The merits and demerits of *Pongamia* as compared to Jatropha are presented below:

- It is a nitrogen fixing tree thus its cultivation improves the soil fertility.
- Being a tree it falls under forest act and cutting and rigorous pruning may pose problems when cultivated commercially on large scale.
- It can thrive in water logged areas where Jatropha cultivation is not suitable. However, no crop can be grown underneath this tree due to shadow. Intercropping with Jatropha is possible.
- The economical yield starts after 6th year as compared to 3rd year in Jatropha.
- The productive life of *Pongamia* is about 70-80 years as compared to 40-50 years of Jatropha.
- There is not much difference in yield /hectare between both the plant species except the plant density.
- It is easier to genetically transform Jatropha for desired traits since it has low gestation period as compared to *Pongamia.*

Distribution

The natural distribution of pongamia is along coast and river banks in India and Burma. Native to the Asian subcontinent, this species has been introduced to humid tropical lowlands in the Philippines, Malaysia, Australia, the Seychelles, the United States and Indonesia. *Pongamia pinnata* belongs to the family Papilionaceae. It is also called *Derris indica & Pongamia globra.* It is a medium sized evergreen tree with a spreading crown and short bole. The tree is planted for shade and is grown as ornamental tree. It is one of the few nitrogen fixing trees producing seeds containing 30-40% oil. The natural distribution is along coasts and river sides, canal banks and open farm lands.

Ecology

Native to humid and subtropical environments, *Pongamia* thrives in areas having an annual rainfall ranging from 500 to 2500mm, in its natural habitat, the maximum temperature ranges from 27 to 38^{0}C and the minimum 1 to 16^{0}C. Mature tree can withstand water logging and slight frost. This species grows to an elevation upto 1200m. but in the Himalayan foot hills is not found above 600m.

Soil

Pongamia grows on moist soil types ranging from stony to sandy clay. It does not do well on dry sands. It is highly tolerant of salinity. It is common along waterways or seashores, with its roots in fresh or salt water. Highest growth rates are observed on well drained soils with assured moisture. Natural reproduction is profuse by seed and common root.

Botanical Features

Pongam (Leguminoceae, sub family Papilionaceae) is a medium sized tree that generally attains a height of about 8m and a trunk diameter of more than 50cm (Fig. 5). The trunk is generally short with thick branches spreading into a dense hemispherical crown of dark green leaves. The bark is thin gray to grayish-brown and yellow on the inside. The tap root is thick and long; lateral roots are numerous and well developed.

Leaf

Leaves are alternate imparipinnate, leaflets opposite, 5 to 9 in number, which are arranged in 2 or 3 pairs and a single terminal leaf. Leaves are ovate to ovate elliptic, shortly acuminate, glabrous bright green in colour, petiole 4.5 cm long (Fig 6). Leaflets are 5-10 cm long, 4-6 cm wide and pointed tip.

Flowers

Flowers are pink, light purple or white. Inflorescence is of auxiliary racemes and shorter than leaves, about 20 cm long. Flowers are one cm across, zygomorphic and style incurved. *Pongamia* starts flowering from the 4th or 5th year of planting. White and purplish flowers in auxiliary racemes appear in April to July.

Fruiting

Out of the two ovules in the ovary, invariably only one will develop into seed. After fertilization the early fertilized ovule suppresses the

Fig. 5. Cultivation of *Pongamia pinnata*

Fig. 6. Morphological features *Pongamia pinnata* plant

subordinate one by the strong sink activity. The abortion of embryo is due to manifestation of sibling rivalry. The yield of fruit varies from 9 to 90 kg per tree for different age trees.

Pods and Seed

Pods are obliquely oblong, 3-6 cm long and 2-3 cm wide thick walled, woody, compressed, indehiscent and usually contain a single seed, pointed at both ends, yellowish grey when ripe and 1 or 2 seeded. Seeds are elliptical, reniform, compressed, reddish brown, fairly hard, 2-3 cm long. The pods ripen from Feb to May in the following year.

Seed Collection and storage

The pods are dried in sun and seeds are extracted by threshing the fruits. They remain viable for about a year when stored with the fruit shell unopened in air-tight containers/stored at 5^0C.*Pongamia* is easily established by direct seeds or by planting nursery-raised seedlings or stump cuttings or 1-2 cm root-collar diameter. In peninsular India, the seeding season is April to June, and the seed yield per tree ranges from about 10 kg to more than 50 kg. There are 1500-1700 seeds per kg seeds, and remain viable for about a year when stored with the fruit shell in open air-tight containers at 5^0C.

Seed Sowing and Germination

Seeds do not require any pretreatment before sowing. But, soaking the seeds in hot water for 15 minutes improves germination percent and vigour. Seeds are sown in seed beds/ polypots/ sand trays with the micropyle facing downwards. Seed germinates within two weeks of sowing. Seedlings attain a height of 25-30 cm in their first growing season.

Transplanting

Transplanting to the field should occur at the beginning of the next rainy season when seedlings are 60 cm in height. Seedlings have large root systems. Soil should be retained around the roots during transplanting. Pits of 30x30x30 cm are appropriate for planting at 3mx3m distance. Soil should be retained around the roots during transplantation. The spacing adopted in avenue planting is about 8 m between plants. in block planting, the spacing can range from 2x2 to

5x5 m. pongam seedlings withstand shade very well and can be interplanted in existing tree stands.

Commercial Uses

Oil

A thick yellow orange to brown oil is extracted from seeds. Yields of 25% of volume are possible using a mechanical expeller. The seeds are largely exploited for extraction of a non edible oil commercially known as 'Karanja oil' which is well organized for its medicinal properties. There is no systematic organized collection of seeds. Mixture seeds consist of 95% kernel and is reported to contain about 27.0% oil. The yield of oil is reported to be about 24 to 26.5% if mechanical expellers are used for the recovery of oil from the kernels, but it is only 18-22% from village crushers. The crude oil is yellow orange to brown in colour which deepens on standing. It has a bitter taste and disagreeable odour, thus it is not considered edible.

The oil has a better taste and a disagreeable aroma, thus it is not considered edible. In India, the oil is used as a fuel for cooking and lamps. The oil is also used as a lubricant, water-paint binder, pesticide, and in soap making and tanning industries. The oil is known to have value in folk medicine for the treatment of rheumatism, as well as human and animal skin diseases. It is effective in enhancing the pigmentation of skin affected by leucoderma or scabies. The oil of pongam is also used as a substitute for diesel.

Medicinal Uses

The oil is known to have value in folk medicine for the treatment of rheumatism, as well as human and animal skin diseases. It is effective in enhancing the pigmentation of skin affected by leucoderma or scabies. Its root, bark, leaf, sap, and flower also have medicinal properties.

Wood

Pongamia is commonly used as fuel wood. Wood is not durable, susceptible to insect attack and tends to split when sown. Thus the wood is not considered a quality timber. The wood is used for cabinet making, cart wheels, posts, agricultural implements, tool handles and combs, etc.

Fodder and Feed

Opinions vary on the usefulness of this species as a fodder. It is reported in some places that the leaves are eaten by cattle and readily consumed by goats. However, in many areas it is not commonly eaten by farm animals. Its fodder value is greatest in arid regions. The press cake, remaining when oil is extracted from the seeds, is used as poultry feed.

Bio-Fertilizers and Bio-Pesticides

Pongamia is a drought resistant, nitrogen fixing leguminous tree. Incorporation of leaves and the press cake into soils improves fertility. Dried leaves are used as an insect repellent in stored grains. The press cake, when applied to the soil, has pesticidal value, particularly against nematodes. It also salt tolerant and to some extent tolerant to slight frost. It is a good shade tree. The shade provided by this tree is said to have cooling effect and is good for health.

Social Uses

Pongamia is often planted in homesteads as a shade or ornamental tree and in avenue plantings along roadsides and canals. When planted as a shade or ornamental tree, branch pruning may be necessary to obtain a trunk of appropriate height. It is a preferred species for controlling soil erosion and binding sand dunes because of its dense network of lateral roots.

Pongam should be grown in full sun or partial shade on well-drained soil. A relatively low maintenance tree once established, is resistant to high winds and drought but is susceptible to freezing temperatures below 30 degrees F. Pongam will show nutritional deficiencies if grown on soil with a pH above 7.5. No pests are of major concern, but caterpillars occasionally cause some defoliation.

Yield and Economics

The plant starts giving yield from the fifth year onwards and the benefits increase over the years and stabilizes in the 10th year. One plant yields 9-90 Kg seed which depends on a vrieties of factors associated. Comparative studies on performance of Jatropha and Pongamia and their physico-chemical characteristic features are given in Table 9.

Table 9: Comparative studies on bio-diesel.

General Characteristics	*Jatropha curcas*	*Pongamia pinnata*
Maturity / Fruiting	3-4 years	7 years
Plants / hectare	1100-2500	156-200
Seed / tree	2 Kg Average (1-4 Kg)	15 Kg Average (10-20 Kg)
Seed yield / hectare	5000 Kg	3500 Kg
Oil yield / hectare	1750 Kg (30-40 %)	1075 Kg (25-30 %)
Tree height	2 m	10 m
Fruit Shell	Too thin	
(easy deshelling)	Too thick	
(deshelling difficult)		
Physico-chemical Characteristics		
Specific gravity (15 0C)	0.918 - 0.923	0.925 - 0.940
Flash point (0C)	191	134
Cetane Index	57 - 62	56.2
Sulphure (%)	0.014	0.02
FFA (%)	5.8 - 7.5	8.3

contd....

21.3 *Jatropha curcas*

Jatropha curcas is large shrub or tree native to the American tropics. However, it is commonly found and utilized throughout the tropical and sub tropical regions of the world. The plant is reported to have been introduced into Asia and Africa by Portuguese as oil-yielding plant. Now it is occurring almost throughout India and Andman Islands in semi-wild condition. Properties like hardiness, fast growth; easy and quick propagation (either by seeds or cuttings) and wide ranging utility have resulted in the spread of Jatropha far beyond its place of origin. The distribution shows that introduction has been most successful in the drier regions of the tropics.

The genus *Jatropha* contains about 175 species in the world. In India 18 species are found scattered in different states. Out of 18 species, *Jatropha curcas* gained prominence because of its added features like best adaptability to different habitats, larger fruits and seeds, high oil yielding, soil conservation capabilities, thrive well as live fence etc.

Jatropha (Physic nut) is a hardy plant, well adapted to arid, semi arid, low rain fall (400 mm) to high rain fall (1200mm) and can be cultivated in soft, rocky, gravelly, sandy, calcareous, saline and sloping soil. This requires low fertility and moisture and can live for 40-50 years in fruit production stage. Different parts of this plant are used in medicine preparation. Bark of this plant is used as raw material for dye, seed cake after oil expeller is used as bio-fertilizer for soil enrichment, seed and leaves are used as insecticides/pesticides. Besides Jatropha cultivation ensures sustenance of ECO^3 (economy, eco-development and employment to rural people). This is a best solution for national energy security and rural economy. Jatropha cultivation also ensures protection of soil by bio-degradable product, checking land erosion, retention of ground water, reduction in environmental pollution, conservation of natural fossil fuel, resources etc. Jatropha plant is adaptable to different climates up to 5000 feet altitude and also thrives well in low fertile and rainfall zones. It is not browsed by animals and has got low gestation period. Plant is resistant to diseases and pests and has low acid value of oil.

The first commercial application of *Jatropha curcas* was reported for the oil, used for soap production, candle preparation and press cake was used as fertilizer (Gubitz et al., 1999). Apart from its oil, it is being used in traditional medicine and veterinary purposes for a long time. The oil is used to treat skin diseases and against many infections.

It also has antiseptic, anti inflammatory, molluscicidal, insecticidal and fungicidal properties (Gubitz et al., 1999).

21.3.1 Jatropha – A Miraculous Plant

- Grows in low fertile and rainfall zones
- Yield is high in irrigated and high rainfall area.
- Hardy plant, easy to establish and grow rapidly.
- Grow easily on marginal, degraded, waste and other low fertility soil.
- Oil yield per ha is highest among TBO,s Plants.
- Can be grown in desert area with drip irrigation.
- Traditionally grown as live fence (Biofencing) to protect crop from browsing animals and strong winds.
- Trees are not tall and seeds are easy to collect as mature after rainy season.
- Easy to propagate through seed, seedlings and cuttings.
- Seed production starts from 1st year 0.4ton/ha to over 5 ton/ha after 3 years.
- Apiculture can be promoted for production of honey and fertilization of flowers.
- Seed cake is rich in nitrogen and used as excellent source of plant nutrients (bio-fertilizer).
- Inter cropping with Jatropha can be done.
- Adaptability is quick.
- Not browsed by animals
- Disease tolerance
- Easy oil extraction & trans-esterification
- Medicinal value of bark, leaf and seed
- Bark used as dye in textile
- Jatropha plantation supports wild life habitat.
- Minimize soil erosion.
- Avoids competition for food crops.
- Does not rely on the use of harsh chemicals and fertilizers.

21.3.2 *Jatropha curcas*- An option for bio-fuel

- Has easy propagation, fast growth, wide distribution, suitable for varying agro-climatic zones. However, need identification of species/varieties and development of agronomical practices for different agro-climatic zones.
- Not damaged by cattle, birds and other animals. Low incidence of diseases and pests, and also other abiotic & biotic stresses have least effects.
- Oil quantity if explored fully can meet almost 65% demand of diesel of the country but need improvement of quality and quantity for better efficiency Cl, engine at par or even better than diesel.
- Has variety of uses viz., medicine in traditional as well as modern therapies, as botanical, as domestic fuel, use in soap industry, as source of biogas, bio-fertilizer & cattle feed.
- No biomass destruction takes place except that of seed which is renewable, but seed waste needs detoxification for alternative uses to safe the environment and ecology.
- Among the various species *J. curcas* has been opted all over the globe, as viable species.

Uncertainties

Jatropha widely occurs and grows in nature in all climatic conditions and has got variety of uses for human being. However, there are certain uncertainties to be looked after for making the plant more economic and viable.

- Bio-engineering for high yielding cultivar
- High yielding cultivars for different location specific and climatic zones
- Minimizing Nox
- Detoxification of cake (curcin).

21.3.3 *Characteristic Features of Jatropha Plant*

Jatropha is called as Ratanjote, Jamalghota, Chandrajote, Kala aranda and Inu. This plant is cosmopolitan in nature, however, it occurs abundantly in Madhya Pradesh, Rajasthan, Haryana and Maharastra It is perennial and cross-pollinated tree. Characteristic features are given below:

21.3.4 *Plant Characters*

Jatropha curcas is a deciduous soft-wooded small multi purpose small tree or shrub, with smooth grey bark which exudes whitish coloured watery latex when cut. It usually attain the height of 3-5 m, however, it can grow even up to 8-10 m in favourable conditions. It is monoecious and protandrous. Young shoots are glandular, tomentose, base is grey and green. Trunk is straight, branched from the ground. Bark is thin and yellowish in colour.

21.3.5 *Leaf*

Leaf is smooth, heart shaped, 4-6 lobed and 10-15 cm (Fig. 7) in length and width, initially light violet later on yellowish green and at maturity it becomes dark green. Leaf fall occurs in the winter season leaving entire plant naked.

21.3.6 *Inflorescence*

Inflorescence is terminal; flowers are unisexual, protogynous with monoecious condition (Fig 8&9). Ratio of male and female flower is 25:1. Female flowers are bigger than male, lesser in number, open 2-3 days after male flower within a plant and these ensure a self-incompatible system. Lesser number of female flowers and inadequate pollination are major causes of low yields. Usually flower initiates in rainy season and bears fruiting in winter, however, this character varies according to geo–climatic condition in India. For early flower giberalic acid (GA) @ 100 ppm can be sprayed. High temperature using summer (40±2^{0}C) effects sex switching from protandry to protogynic (signal transudation) and capsule with few unfilled seed maturing in late June-July.

21.3.7 *Fruit Set and Seed Formation*

Jatropha plant takes 2-3 years to commence fruiting and another 2 years to come to stage of full bearing. One fruit bears 3 seeds. A normal size of seed measure 17-20 mm and total number of seed of this size is 1500-1700 in one kg. and light weight and small sized seed may number upto 2200-2400 seed / Kg.Healthy seeds give 90% germination. In case seed is kept for long period germination percentage goes down. Jatropha can yield fruit in first year also but harvested yield comes in 3rd year. However, economic yield time is 5th year of plantation.

cultivation in India came in big way in 2003. With the wave of bio-diesel production from Jatropha Indian Scientists / farmers are running after and even Govt. / Non Govt. departments have chalked out plan for plantation in a large area without having standard agro-technological practices. However, fragmented information on Jatropha agro-technology is given by various worker.

21.3.9 Ecological Requirement

Jatropha exhibits a wide environmental tolerance; it is found in seasonally dry tropics as well as equatorial regions and is well adapted for cultivation within the vast areas of marginal and degraded lands in semi-arid and arid tropics. It occurs mainly at lower altitudes (0-500 m) in areas with average annual temperatures well above 20^{0}C but can grow at higher altitudes (1500m) and tolerates slight frost, however, very cold areas do not suit to this plant.

21.3.10 Rain

Jatropha is hardy to dry weather conditions and can be grown over a wide range of arid or semi-arid climatic conditions. Jatropha grows well in low rainfall conditions (400 to 750 mm). It can also tolerate high rainfall (up to 1200 mm) conditions and can stand long periods of drought by shedding most of its leaves to ruduce the transpiration loss. Rain plays an important role in inducing flowering.

21.3.11 Climate

It tolerates annual temperature range of 18-28^{0}C even higher ranges but it can not tolerate very harsh winter or frost. For the emergence of seeds, hot and humid climate is preferred. The flowering is induced in rainy seasons and bears fruits in winter. The foliage drops with dip in temperature during winter and with rise it starts blooming.Jatropha has very strong adaptive mechanism to sustain variable climates. It can tolerate extremes of temperature but not the frost. The frost damages plants whereas, high temperatures adversely affect yield. The germplasm should be screened to isolate lines which can withstand high temperature.

21.3.12 Waste Land Scenario in India Cultivation: Types

Out of total 329 million ha land area of the country 60 million ha area is a wasteland. Jatropha can grow on lands unsuitable for economically viable agriculture. The suitability of Jatropha to low fertility wasteland situation besides other cogent reasons may be raised on manageable wasteland on priority to establish plantations. There are 6 types of waste lands available in Country (Table 10). The first five types of wasteland can be used for plantation under Jatropha with some management and yield sacrifice. The land with scrub (15.1 mill. Ha) out side forest and degraded forest with scrub (10.9 mill. Ha) should begin simultaneously.

Table 10: Waste land for Jatropha plantation.

S. No.	Land type	Area in million ha
1.	Shallow/medium ravenous	1.5
2.	Land with scrub	15.1
3.	Land without scrub	3.7
4.	Saline / alkaline	0.4
5.	Shifting cultivation	3.5
6.	Degraded forest-scrub	10.9
	Total waste Land for Jatropha	31.1

The land holding with farmers constitute 170 million ha under cultivation which is being slowly brought under urbanization and development of other projects hence; the land under cultivation is reducing. It is a serious concern and threat to food security. Therefore no land under crop cultivation be brought under Jatropha plantation. The last three two offer land close to culturable land, which totals to 42 mill.ha may be put to plantation (Table 11).

Table 11: Land may be available for Jatropha plantation.

S. No.	Land type	Area (Million ha)
1.	Land under cultivation	170
2.	Land suitable for mixed cropping	25
3.	Land under field boundaries	25
4.	Pvt. Cultivated Land likely to be under Jatropha	5
5.	Degraded & cultivated Land likely to be under Jatropha	12

21.3.13 Land Required on Blending Basis

As per blending norms at 5%, 10% and 20% total land required is 1.185, 2.37 and 4.75 mha respectively, however, Indian Railways alone require 52000, 104000, and 208000 ha respectively (Table 12).

Table 12: Land required on blending basis.

Blending%	Indian Railway (ha)	India (Million ha)
5%	52000 ha	1.185
10%	104000 ha	2.37
20%	208000 ha	4.75

21.3.14 Land Preparation

Though plantation can be done without any cleaning activities, however, it is advisable to partly clean up the area. Tall trees can be left as such and bushy plants have to be cleaned.

21.3.15 Soil Reclamation

Jatropha curcas is a plant that can grow almost any where, even on soft, rocky, gravelly, sandy, calcareous, saline and sloping soils. It has low fertility and moisture demand. To combat phosphate deficiency it avails of the symbiosis with root fungi (Mycorrhiza). The crop is undemanding in soil type and even does not require tillage. It can grow even in crevices of rocks. It does not thrive in wetland conditions. It has been experienced that the land where the soil depth is less than 15 cm plantation of Jatropha should not be taken up as a commercial crop as the expected returns shall not be available form such soil. The leaves shed in winter months and form mulch around the base of the plant. The organic matter from shed leaves has been observed to enhance earthworm activity in soil around the root-zone of the plants, which is a fair indicator for improvement in microfauna and fertility and texture of soil.

21.3.16 Use of Mycorrhiza

Mycorrhiza is a symbiotic, non-pathogenic, permanent association between the roots of land plants and a group of fungi. Mycorrhizae are essential soil organisms that are more than 400 million years old and played a key role in allowing plants to colonize land plants. The most common among these are AM (*Arbuscular mycorrhiza*) fungi benefiting 90% of the land plants (Gaur, 2006). They provide extended arms to the plant root helping explore more soil nutrients and thus providing a selective advantage to the plant in terms of general health and better survivability. Some of the benefits and uniqueness of this group of fungi include:

- It offers up to 50% reduction in the phosphorus fertilizer application.
- It allows better uptake of nutrients like phosphorus and immobile trace elements like Zinc. Cobalt. Magnesium, Iron, Copper, Molybdenum, etc.
- It offers tolerance against range of soil stresses like heavey metal toxicity, salinity, drought and high soil temperatures.
- Higher resistance to various soil-borne and root-borne pathogens thus it becomes a potential bio-control agent.
- It helps in soil conservation and soil structure stabilization, thus restoring land productivity.

Jatropha is itself a very hardy plant and with its known naturally existing association with fungi, this combination will not only give higher yields with respect to seed production but will also give higher vegetative biomass. A technology developed by TERI, now ensures means of availability of this bio-fertilizer on a mass scale commercially.

21.3.17 Varieties of Jatropha

Since Jatropha is wild plant and occur in different geo-climatic condition of the country which obviously envisages variation in production and productivity of seed and oil contents. Jatropha came in big way since 2003 in India. So far there is no any identified variety for a particular region which can be recommended. However, exploration of the variety occurring in nature is in progress and development of high yield variety through conventional breeding and biotechnological approaches are also in progress. There is need to develop / screen suitable varieties of higher seed yield per unit with high quantity and good quality oil for different agro-climatic conditions. The non-availability of improved variety itself is major constraint to begin with the plantation.

Two varieties viz., Nicaragua and Mexican types are reported from western world. Mexican types are non toxic and the Nicaraguan types are poisonous type. The center of diversity of land races and ecotypes are Central and South America. In Madagascar *J. mahafaensis* (2n=22) is commonly cultivated. In India so far local varieties are cultivated, however, Chhatrapati (SDAU-1), Urlikanchan, Hansraj and Sardar Krishinagar (SKN-B), SKN-1, JH4 and IST 1Big have been developed by Sardar Krishinagar Dantewala Agricultural University and these varieties are suitable for Rajasthan and Gujarat. Tamil Nadu Agriculture University has developed two varieties i.e. TNMC-6,7 and 24. Defence Agricultural Research Laboratory, Haldwani has collected various strains from different geo-climatic conditions and after analysis from IICT, Hyderabad two strains viz., DARL-1 and DARL-2 have been identified promising with oil content 34.4 and 36.5 % respectively. Unique traits viz., high oil content, high yield, more drought tolerance character, resistant to insect-pest and diseases etc are being transferred by crossing with specific character to develop high yielding varieties/ hybrids. FCRI, INAO, Mettespalayarn, JNKV, Jabalpur, MPUA & T, Udaipur, CSK, HPKV, Palampur, NAU, Navsari, Gujarat, CCS HAU, Hissar (Kausik, 2006) and various other universities, Institutes, CSAUT, Kanpur, SKAUST, Jammu, TFRI, Jabalpur, NRC, Janshi, NDUAT, Faizabad and NGO's are working on varietals improvement.

The major characteristic features of Jatropha to be considered are production and productivity potential and oil content to make the bio-diesel production viable and economic. These two characters vary as per geo-climatic condition, soil texture, irrigation facilities and rainfed condition. Mexican varietie contains highest oil percentage (58-60%) followed by Nigerian variety (50-55%) as compared to Indian variety (30-55 %). In India there is big variation in oil content. After collection of Jatropha germplasm from different climatic conditions and getting analyzed from Indian Institute of Chemical Technology, Hyderabad ,oil percentage varied from 20- 36.5% which needs more emphasis on varietal improvement and its stability (Table 14).

Table 13: variation of oil yield of *j. curcas* world wide.

Countries	Oil percentage
India	30-58%
Austria	45-55%
Bankok	40-52%
USA	35-55%
Nigeria	50-55%
Mexico	58-60%

Table 14: Oil analysis of different germplasm of *J. curcas* (Indian Institute of Chemical Technology, Hyderabad).

Sl.No	Source of Seed Germplasm	Oil Content %
1.	Bio-diesel Society, Khammam (AP)	31.6
2.	NIRD, Hyderabad	20.0
3.	Jatropha Bio-diesel Dev-Orgn, Khammam	31.9
4.	SKNAU, Gujarat	25.5
5.	Forest Deptt, Pandra Road (CG)	31.3
6.	GBPUA&T, Pantnagar	24.7
7.	DARL-1 Pithoragarh- (Uttarakhand)	34.4
8.	DARL-2 Pithoragarh- (Uttarakhand)	36.5
9.	DARL-5 Pithoragarh- (Uttarakhand)	21.9
10.	DARL-7 Pithoragarh- (Uttarakhand)	28.6
11.	DARL-10 Pithoragarh- (Uttarakhand)	28.3
12.	DRLT-1	21.9
13.	Medors Biotech New Delhi	30.5

21.3.18 Varietal / Crop Improvement

Plant breeders have long since developed crops with oils suited to different specific purposes. These practices are, however, very time consuming, taking several years to be perfect. Advances in plant molecular biology have revealed the biochemical pathway by which plants make oil and influence that different fatty acids have an oil characteristics. Genetic modification now makes it possible to improve the composition and properties of oil from different plants far more quickly and precisely than with traditional breeding techniques. Systematic work would need time for improvement, evaluation and release of variety, sex switching leading to change in proportion of male vs female flower, allelopathic effects and thermo-insensivity. Considering the potential application of recombinant DNA technology in modern agriculture, it is essential to briefly discuss lipid metabolism at molecular level with major emphasis on storage oil in plants as it helps in designer oil production. Keeping in view of genetic improvement introduction of quality germplasm, selection of germplasm geo-climate wise and molecular manipulation may be considered rapidly to coupe up with energy security. Introduction of quality germplasm from potential countries to broaden genetic base for yield enhancement has to be done.

Fig. 11. Seed treatment before sowing

Fig. 12. Quality seed of Jatropha

Fig. 13. Root development in seed sown plant

21.3.23 Nursery Raising

21.3.23.1 Polybag Nursery

For transportation purpose nursery sowing in preferably black poly bag (Fig 14, 15 & 16) is good, however, it is cost effective and labour intensive as compared to sowing in soil bed. The polythene bags of 10x20 cm and 15x 25 sizes have been found suitable. The saplings in this size can be maintained for the period of three months. The size of poly-bags may be decided as per requirement.

The overnight pre soaked seeds germinate in 6-10 days in hot humid weather, whereas the process continues for 10-15 days. It takes more time in low temperature conditions. Saplings from seeds develop a typical taproot and four lateral roots. The nursery should be irrigated as and when needed. The areas where high temperature prevail (40±1) for longer duration locally available materials like grass straw etc. can be used to create shade till rainy season. The saplings are normally ready for transfer 50-60 days after seeding. In an experiment overnight soaked seed in normal water gave 80-85% (Table 15) germination as compared to un soaked (70-75%) and soaked in warm water (65-70%).

Table 15: Effect of soaking on germination.

Seed Treatment	Days to 50% germination	Total germination (%)
Overnight socked	10	80-85
Socked in warm water-4-5 hr	09	65-70
Un soaked	11	70-75

Fig. 14. Poly bag nursery raising

Fig. 15. View of poly bag nursery

Fig. 16. View of poly bag nursery

21.3.23.2 Soil Beds Nursery

The large scale plantation can be done in soil beds (Fig 20-25) to raise saplings for transplanting. Nursery raised on soil beds can be taken out bare root for transportation even to a long distances. The transportation should be done on rainy days or packing in water soaked media. Soil bed is prepared in field by raising soil about 3-4 inches to avoid water flood/logging in soil bed. Size of soil bed measuring 1.5 X 10 meter is appropriate (Fig 17&18) and convenient for intercultural operation. After sowing of seed light shower irrigation in the evening or early morning is required till germination (Fig 19). Soil bed nursery can also be transferred in mixture filled poly bags in case it is required. The seeds start emerging within 6-7 days. The plantations from nursery raised saplings have shown high rate of survival and better establishment. The saplings at 3-4 leaf stage are ideal for transplanting and it grows fast.

Fig. 17. Soil bed for nursery raising

Fig. 18. Soil bed nursery- germination

Fig. 19. Soil bed nursery -germinated

Fig. 20. Soil bed nursery-large scale

Fig. 21. Inspection-Soil bed nursery

21.3.23. 3 Stem Cutting

Jatropha is pruned after maturing of stem/branches. These stem should be cut into pieces measuring 30-40 cm (Fig 26) depending on the length of branch and planting should be done in soil bed as given below:

Cutting size	-	30 -40 cm
Buds / cutting	-	8-10 no.
Spacing	-	10 X 10 cm
Soil mixture	-	Soil mixed with compost
Transplantation	-	90 days old

Cuttings can be raised in poly bags also but required watering is necessary to sustain plants for sprouting of roots and leaves. Six months to one year old shoots are selected for cuttings, however, at the time of pruning total branches up to a height of 45-60 cm above the soil are cut and mature branches are taken for cuttings. The thick strong shoots of 30 cm long with 4-5 buds preferably taken from the middle of the branch are the best suited material for propagation as they give nearly 75-85% rooting in case required watering is done. Rooting percentage may be higher if branches are fully matured. The cuttings are planted in raised beds of 1.5x 10 m size. The soil is mixed with well rotten farm yard manure. The cuttings are planted closely with the spacing of 10x10 cm to accommodate more numbers in a less area. The cut ends of the stem cuttings are dipped in 0.3% benlate or wet cerason. Dipping of cut ends in cow dung solution gives good response. The beds are watered regularly.

Fig. 26. Nursery raised through stem cutting

Fig. 27. Bundles of Jatropha nursery

Fig. 28. Uprooting and bundling of Jatropha nursery

Fig. 29. Loading of Jatropha seedling in open truck

Fig. 30. Loading of Jatropha seedling in open truck

21.3.24 Spacing

For plantation plant to plant & row to row distance may be decided according to soil type, rain fall pattern and irrigation facilities available. The 2x2 m spacing (Fig. 31 & 32)accommodates 2500 plants per ha under irrigated or partially irrigated conditions. In south Indian conditions 2 x 1.5 m spacing (3333plants) is found to be ideal. In rain fed wasteland, high density planting with a spacing of 2x1 m or 1.5 x 1.5 m accommodates 5000 or 4444 plants per hectare respectively. Satisfactory planting densities are 2x2m, 2x2.5 m 2.5x 2.5 m , 3x3m and 4 x 3m. This is equivalent to the crop densities of 2500, 2000, 1600, 1111, and 833 plants/ha respectively (Table 16).

Table. 16: Plant density and land types.

Distance (m)	No. of Plants/ha	Types of land/Irrigation
2 X 1.5 m	3333	Non irrigated / low rainfed area
2 X 2 m	2500	Non irrigated/ rain fed area
2 x 2.5 m	2000	Semi irrigated
3 X 2 m	1666	Irrigated areas
3 X 3 m	1111	Inter cropping
4 X 3 m	833	For tractor operation & grass collection
0.5 X 0.5 m	4000	For fencing purpose

Fig. 31. Row to row spacing

Fig. 32. Plant spacing 2x2 m row to row and plant to plant

21.3.25 Manure and Fertilizers

Chatturvedi (2006) reported that by processing Jatropha crop of 1 million ha, apart from getting 20 lakhs tonnes of bio-diesel, 40 lakhs tonnes of premium quality of bio-manure would be received for use in agriculture sector. This bio-manure (possessing 6% N) besides providing much needed major and micro nutrients to the soils will also help in preserving soil humus, so much necessary to preserve soil quality which is getting imbalanced due to persistent use of chemical fertilizers. Even by taking a nominal cost of rs. 2/- per kg for this bio-manure, the value of bio manure so produced will be about Rs. 800 crore annually.

Although Jatropha is adapted to low fertility site and alkaline soils, better yield is obtained on poor quality soils if fertilizers with small amounts of calcium, magnesium and sulphur are used. Mycorrhizal associations have been observed with Jatropha. In general application of super phosphate@ 150 kg/ha and alternated with one dose of 40:100:40 kg/ha NPK at half yearly intervals is reported to improve the yield. The application should coincide with rainy season or followed by proper irrigation immediately after application of fertilizers. From fourth year onwards 10% extra super phosphate should be added to the above dose.Soil fertilization with different fertilizers is recommended (Table 17).

Table 17: Soil fertilization and doses.

Fertilizer	Types	Dose/plant
Organic	FYM	5 kg/plant
Inorganic	Urea	20g
	Superphosphate	120g
	Muret of Potash	16g
Biofertilzers	AZO	10g
	VAM	50g

Before transplantation 1 Kg. FYM/Vermicompost, 10 g Urea , 20 g phosphate and 5 g Potash are mixed with soil in ¾ of pits and after that plantation is done. This quantity of fertilizer ensures good performance of plants, however, standardization of fertilizer dose is still to be standardized based on soil structure, location and geoclimatic condition.

21.3.26 Transplantation

For better establishment of saplings, monsoon season should be preferred for planting (Jun-Sept). The pits are filled with 500 g farm yard manure (FYM) and 100g neem cake or vermin compost to ensure profuse rooting in nutrient deficient soils. Transplanting in the field should be done preferably in the evening. The saplings removed from nursery should be kept in shady place to avoid wilting. Plants transplanted with bare naked root has to be kept in soil where shady place is available and depending upon the requirements watering is done if by any cause transplantation is delayed. Air is taken during uprooting, bandling and transportation of Jatropha seedling (Fig, 28,29,30).

21.3.27 Hedge Rows

The spacing for hedgerows is 15 to 25 cm in one or two rows. Thus there will be between 4,000 to 6,700 plants per km for a single hedge row and double that when two rows are planted. Dense hedge rows do not allow wild animals or cattle's to destroy the crop and it acts as fencing known as bio-fencing.

21.3.28 Plant Geometry

The major criteria for planting density and geometry (square, rectangle and alternate) are based on the soil type and availability of facilities for irrigation (Fig 33, 34 & 35). The alternate planting geometry guarantees the availability of sunlight to every plant for a longer period of time with less of shadow affect.

Fig. 33. Plant Geometry

Fig. 34. Diagonal rows

Fig. 35. Plant Geometry -Diagonal rows

21.3.29 Digging of Pits

Jatropha plantation is done in appropriate size of pits. The site preparation, cleaning and leveling of land for plantation can be used to dig the pits without tillage. Pits of standard sizes are dug initially, based on the slope of land, availability of water and quality of soil. The planting density in fertile soils should be lower than in soils with low fertility. The pits should be dug in with proper layout to mechanically managed plantations. The pits size of 30x30x30 cm is ideal (Fig 36&37) for plantation in soils fairly rich in nutrients.

Fig. 36. Pit digging

Fig. 37. Pit size

21.3.30 Irrigation

The seedlings require irrigation especially during the first 2-3 months of planting (Fig 38&39). The requirement of water is essential as per soil and climate conditions. The stage of fertigation should be matched with the time of irrigation. After transplantation pits should be filled with soil and watering is done. To establish the plant fortnightly watering / irrigation for two month is required.

Fig. 38. Irrigation of Plant

Fig. 39. Irrigation of Plant

21.3.31 Pruning and Plant Canopy Management (PPCM)

Pruning is done in such a way that a plant has to have 15-25 branches so as to obtain optimum production and productivity. Crop architecture (Fig 40&41) plays an important role in a plant like Jatropha where proper pruning will help in producing more branches, (Fig 42) healthy inflorescence to enhance good fruit set and ultimately the yield. The pruning of terminals is essential at a height of 45-60cm at six months age to induce laterals. Likewise the secondary and tertiary branches are to be pruned at the end of second year. During the second year each side branch should be pruned up to $2/3^{rd}$ top portion and retaining $1/3^{rd}$ of branch on the plant, Periodical pruning can be done depending upon the vegetative growth of the plants.

To ensure maximum branching pruning should be done when the tree sheds leaves and enters into a period of dormancy preferably during winter seasons. The trees are kept in short height so as to manage cultural operation, pesticide application and fruit harvesting manually. It also facilitates picking of mature capsules. The canopy management is advisable in trees with terminal bearing. The plant type with branch in every leaf axel should not be pruned vigorously. Once in 10 years the entire plant has to be cut to ground level leaving 45 cm stump for rejuvenation. The re-growth is quick and starts yielding in about a year. It induces new growth and helps stabilize yield.

It has been observed that after 1st pruning average number of branches is 3.43 with range of 1-6 and after 2nd pruning average number of branch is 9.5 with range of 3-32. However, maximum number of branches have been observed 65 in few plants. To increase the number of branches 3rd pruning may be done at short interval so as to have more that 25 branches per plant (Table 18).

Table 18: Pruning Performance.

At 6 month interval	1st Pruning	2nd Pruning
No of Branches	1-6	3-32
Av. No. of Branches	3.43	9.5

Fig. 40. Crop architecture

Fig. 41. Plant canopy

Fig. 42. Pruning of plant

21.3.32 Inter Culture Operations

The standard cultural practices involve timely weeding (Fig 43), proper fertilization, surface ploughing and pruning. The field should be kept free from weed at all times. Around 3-4 weeding in the initial period is enough to keep the field free from weeds until the crop crosses the height of weed / grass. Light harrowing is beneficial during the early growth stage. The spacing of 2x2 m initially allows tractor mounted implements for various operations but the later stages of plant growth suggests to opt for 2x2.5 m in partially irrigated conditions. Power Triller and Tractor are most suitable for Inter cultural operation in big plantation done with a plant to plant and row to row distance of either 3X3 or 3x2 m in the field.

Fig. 43. Inter culture operation

21.3.33 Weeding

The mechanical weeding with optimal space for movement of tractors (Fig 44) can be done using rotavator. It clears weed in wider area away from plants. The chemical weeding can be done in small manageable plants with weedicides by covering plants with polythene bags using paraquat @ 2 ml/lit of water in early stage of the plant. Basin should be kept free from weeds. The hand weeding or hoeing should be practiced to remove the weeds growing close to plants. For good plant growth weeding is must in 1st year of plantation otherwise grasses dominate over Jatropha plants and causes retarted growth.

Fig. 44. Inter culture operation

21.3.34 Flowering

Plant flowers during the wet (rainy) season and two flowering peaks are often observed. In humid regions flowering occurs throughout the year. Concentrated flowering occurs during July to August. The plant produces yellowish green flowers in racemose inflorescences with dichasial cyme pattern. Numerically, 1-5 female flowers and 25 to 93 male flowers are produced per inflorescences. Female flowers are quite similar to male flower in shape but are relatively larger. The flowering depends on the location and agro-climatic conditions. Inflorescence in Jatropha is terminal having unisexual flowers with male and female ratio upto 25:1. The inflorescence emerges on tip of the stem. Usually flower initiates in rainy season and bears fruiting in winter, however, this character varies according to geo -climatic condition in India. For early flowering giberalic acid (GA) @ 100 ppm can be sprayed. Inflorescences normally exhibit protandry, however, high temperature during summer (40±2^{0}C) effects sex switching from protandry to protogyny due to strong signal transudation and thus few capsules are with unfilled seed. Normally Jatropha flower only once in a year in northern India, however, in Tamil Nadu fruiting occurs almost throughout the year. Few lines are bimodal, which flowers two times in a year. The flowering is mostly continuous in such types in presence of optimum moisture. Lesser number of female flowers and inadequate pollination are a measure cause of low yields. Establishment of apiary in Jatropha plantation can increase production and productivity potential besides honey and wax production in addition to pollination promotion in other crops /plants.

21.3.35 Fruiting

Jatropha takes 2-3 years to commence fruiting and another 2 year or more to come to stage of full bearing. Few germplasm starts fruiting in first year of plantation, however, percentage of fruiting plant is about 35-45% with less production. In case irrigation is perfect fruiting percentage increases. Fruits are grey-brown capsule (Fig 45&46), 4 cm long and generally tri-halved, each comprised of one seed. Seeds are black, about two cm long and one cm thick. Generally fruits are matured by September-October. The seeds mature in three months after flowering. The flowers are pollinated by insects especially by honey bees. When the fruits begin to open, the seeds inside are mature. Seeds cannot be stored for long time as these loose viability within six months. Generally it has nearly 400 - 422 fruits/kg and with 1400-1700 seeds/ kg with average weight of 0.63g.

Proportionate weight of seed kernel, seed coat & fruit husk is given below:

Seed weight	-	60.00%
Seed coat weight	-	23.40%
Seed Kernel weight	-	36.60%
Fruit husk weight	-	40.00%
Fruit husk and Seed coat	-	63.40%

21.3.37 Seed Removal

The harvested fruits should be spread in a single layer on the hard or cemented threshing floor for drying. The fruits should not be dried under the sun. After properly drying, the seeds can be removed manually or a seed decorticator can be used. G.B.Pant University of Agriculture & Technology, Pantnagar has developed mechanical decorticator with a capacity of 373 Kg/hr. IIT, Delhi has developed a power-operated decorticator for Jatropha seeds with a capacity of 150 kg seeds/hr. NOVOD has established a technology for individual farmers for decorticating Jatropha seed with a capacity of 20 kg/hr. Defence Institute of Bio Energy Research, Haldwani has installed desheller with a capacity of 500 Kg/ hr.at its project site Secunderabad in collaboration with CSMCRI, Bhavnagar

Hand operated decorticator (capacity 20 kg pods/h) and power operated decorticator (capacity 40 kg pods/h) have been developed by TNAU Coimbatore. Single fruit bears usually 3 seeds, however, 1,2 and rarely 4 seeds also occur.

21.3.38 Yield

The seed yield is an expression of the combined effect of variety of intercultural operations, application of inputs etc. and beyond human control are edaphic, climatic and rhizospheric (especially in plantation crops) variations. The yield predicted from Jatropha ranges from one to five tons per ha whereas, 2.5 ton is achievable and yield target of 5 ton can be realized on uncultivable wastelands with optimal input including field bunds. The yield increases till the sixth year & stabilizes thereafter. It is mainly influenced by the planting material and management practices. Estimated year wise yield is given in Table 19.

Table 19: Estimated Yield of Jatropha.

Year of plantation	Yield/plant (g)	Yield/acre (kg)	Yield/ha (kg)
1	100	100	250
2	400	400	1000
3	1000	1000	2500
4	2000	2000	5000
5	2500	2500	6250
6	3000	3000	7500

Economic yield: Yield in rain fed condition - 200g to 2.5 kg/plant
In irrigated condition - 500g to 5.0 kg/plant

21.3.39 Intercropping

Since row to row and plant to plant distances of Jatropha variety as per geo-climatic condition, soil structure, terrain etc, however, 2 x 2, 2 x 3, 3 x 3 and 3 x 4m distances are in common practice. In 2x2 and 3x3 m distances inter cropping can be practiced till 3rd year of plantation when the plant come in fruit production stage and attains its maximum height leaving less area in between. However, in 3x3 and 4x3 m distances intercropping is practiced for long period because of maximum space in between row and plant. Usually location specific low volume and high value crops are chosen for intercropping. Pigeon pea, Chilli, Bean, Urd, Ground nut, turmeric, sunflower, seasum, black gram, Green gram, Cow pea, Tomato, Onion, Cucumber, Muskmelon ,Lentil, Linseed, Banana, Pumpkin, Moringa and some flower plants like Mary gold and grasses viz., Andropogan, Guinea, Napier and Tongo etc. can be grown successfully (Fig 48&49) which remunerate additional income. Experimental studies on agronomical practices for intercropping is being carried out in different institutions/ Universities, however, it needs perfect standardization of practices (SOPs) DIBER.

Fig. 48. Inter cropping with Chilli

Fig. 49. Inter cropping with Chilli

21.3.40 Plant Protection

21.3.40.1 Insect Pests

Jatropha is least browsed by cattle, wild animals and birds, however, there is report on incidence of different insects and disease at different stages of the plant. From the literature reviews and survey insect-pest of minor & major importance are listed in Table 20.

Table 20: List of Insect Pest of Jatropha.

S. No	Genus/ Species	Common name	Order: Family	Nature of damage
1.	*Calidea dregei*	Blue bug	Heteroptera	sucks on fruits
2.	*Nezara viridula* L.	Green Stink Bug	Heteroptera: Pentatomidae	sucks on fruits
3.	*Pachycoris klugii* Burmester	Bug	Heteroptera Pentatomidae	Sucks on inflorescence & fruit sap
4.	*Chrysocoris stolii*	Pentatomid Bug	Heteroptera Pentatomidae	Feeds on inflorescence and suck sap of young fruits
5.	*Scutellera nobilis* F.	Pentatomid Bug Spotted Bug	Heteroptera: Scutelleridae	Feeds on flowers and fruits
6.	*Pinnaspis strachani*	Cushion scale	Heteroptera: Diaspididae	Inflorescence and leaves

7	*Oedaleus senegalensis*	Locust	Orthoptera	Feeds on leaves & inflorescence
8.	*Stomphastis (Acrocerops) thraustica* Meyerica	Blister miner	Lepidoptera: Garcilariidae	Mines leaves
9.	*Spodoptera litura* F.	Tobacco caterpiollar	Lepidoptera: Noctuidae	Leaf & flower
10.	*Achoea janata* L.	Semi looper	Lepidoptera: Noctuidae	Feeds on leaves
11.	*Agonosoma trilineatum*	Seed feeding beetle	Coleoptera: Scutelleridae	Feeds on seed
12.	*Oxycetonia versicolor F.*	Beetle	Coleoptera: Scarabaedae	Feeds on inflorescence
13.	*Inderbela*	Bark eating	Copleoptera	
14.	*Pempelia morosalis*	Shoot and Capsule Bora	Lapidoptera	Webbing and feed on inflorescence & bores into capsule
15.	*Selebria morosus*	Shoot and capsule borer	Lepidoptera Pyralidae	Brown caterpillar webs the terminal leaves with silken threads

Meshram and Joshi (1994) reported *Spodoptera litura* as a pest of *Jatropha curcas* for the first time. Shanker and Dhyani (2006) Reported few insect pest of *Jatropha curcas.* Fifteen insect species of order Heteroptera extract nutrients from Jatropha and key pest in Nicaragua is *Pachycoris klugii* occurring at a density of 1234 to 3455 insects per ha(Grimm,1996, Grimm and Guharay (1998) and Grimm and Maes,1997). A global list of phytophagous insects consisting of 60 species in 21 families and four orders has been compiled in Australia where Jatropha is considered as a weed. Seed feeding insect *Agonsoma trilineatum* of family Scutelleridae is a serious pest of Jatropha (Smith and Heard, 2003). *Scutellera nobilis* causes (Fig. 51) flower fall. *Pempelia morosalis* causes webbing and feed on inflorescence and later on bores into the capsule. Arif et al. (2007) reported 3-11 larvae of leaf miner (Fig. 50) per leaf in nursery as well as in main plantation during the month of August and September in Secunderabad. Spray with Metasystox / endosulfan @ 0.015% or 0.02% Dimethoate as a prophylactic application avoids insect incidence, however, at the time of infestation the spray of these insecticides at fortnight interval controls infestation.

Attack of mites on leaves are reported and it causes crinkling on surface of leaves. They can be controlled by the various herbal pesticides e.g. mixtures of vitex, neem, aloe, Calatropis or Rogor @ 2 ml/lit. of water. Opmite @ 2 ml/lit and wettable sulphur are being tested against mite.

Fig. 50. Fruit sucking insect

Fig. 51. Population of Pentatomid bug

21.3.40.2 Diseases

The collar rot (caused by *Myerophomina phaseolina*) may be the problem in beginning and can be controlled with 0.2% Copper Oxy Chloride (COC) or 1% Bordeaux drenching. It may become a serious problem in some areas during monoculture under irrigated conditions. It is caused by *Macrophomina phaseolina* or *Rhizoctonica bataticola*. The rotting at adult stage has been observed in the soils saturated with moisture for a long period of time. The *Jatropha curcas* leaf spots Cercospora are reported to be associated with this species. The rot can be controlled by application of 1% Bordeaux drenching. Minor disease such as root rot (*Fusarium moniliforme*), Damping off (*Phytopthora pithium*) and leaf spots are reported to be caused by *Helminthosporium tetramera, Alternaria* sp. and *Pestalotiopsis* sp. In nursery damping off and in field root rot are common diseases. Leaf spot can be controlled to some extent by sprarying of Butox (0.2%) and root rot by Thiram (0.2%). Dry root rot caused by *Macrophomina phaseolina* and powdery mildew (*Oidiopsis* sp.) are also reported on Jatropha.

21.3.41 Seed Storage

The seed of Jatropha stored in ambient conditions maintains viability for 7-8 months. The long storage affects seed viability beyond eight months. Therefore, the seed being used for plantation purposes is kept at low temperature without loosing viability and effective emergence.

The oil industry requires continuous supply of raw material for oil extraction and esterification. The seeds/kernel containing the oil must be properly stored and prepared for extraction, to maintain high quality in the final product. The long storage of grain is reported to affect oil quality and quantity hence long storage should be avoided. The drying of seeds up to 4 - 8% moisture enhances storability and higher moisture causes incidences of stored grain pests and diseases. It should be maintained to enhance storability.

21.3.42 Technology Status of Jatropha

A large number of oil yielding plants have been identified in various parts of India. *Jatropha* has been selected for focused development in the country due to its short gestation period. While the Jatropha seeds are used for oil extraction, other parts of the plant i.e. leaves, bark etc. can be used for developing organic dyes, medicines, bio-fertilizers. Jatropha cultivation on forest waste land has been projected as an important oil seed plant (Joshi, 2005) and an opportunity for socio-economic development and cleaner environment (Kantwal and Soni, 2003). Chhatisgarh, Gujarat, Rajasthan, Andhra Pradesh and Karnataka are leading states in India for bio-diesel production from Jatropha and *Pongamia* (Anonymous, 2006).

The *Jatropha curcas* is available in almost all the states in scattered manner under traditional plantation with low productivity of seed and oil content. Countrywide search is being made to identify superior germplasm. *Jatropha, being* a member of Euphorbiaceae family has a high adaptability to thrive under a wide range of physiographic and climatic conditions in India. It grows in almost all parts of the country up to an altitude of 5000 feet. However, frosty weather prevents it from flowering.*Jatropha curcas* has been extensively explored for its taxonomy, morphology, geographical distribution, ecology, propagation, cultivation and many suitable agro technologies. It is a broad leaf evergreen plant native to American species that was introduced through out the Caribbean, Africa and Asia by Portuguese in 15th Century. *Jatropha* thrived throughout the pan tropics because it can grow on very marginal soils and can withstand long periods of drought. Being poisonous to animals and insects, it makes an excellence living fence or protective hedge row around gardens and farms.

Center of excellence on Jatropha at TNAU, Coimbatore is also developing cultivation and management practices (Parmathma et al., 2006). Comprehensive documents on Jatropha agro- technology for training scheme have been prepared by GBPUA&T, Pantnagar (Bhattacharya et al, 2005 and Shukla and Negi, 2005). Present status and future prospects of large scale Jatropha plantation in India and waste land development with Mycorrhiza technology has been presented by TERI (Bhojvaid, 2006). Cultivation of Jatropha as medicinal plant crop and studies on oil content, physico-chemical properties, fatty acid composition and energy value have been conducted by NBRI (Banerjee et al., 1985 and Dutta and Pandey, 1993).

The Indian Railways have a plan to replace 10% of its total petro-diesel requirement by bio-diesel. The Ministry in technical collaboration with Indian Oil Corporation Ltd. has started a pilot project by raising Jatropha along the railway track. Bharat Petroleum Corporation Ltd (BPCL), Nandan Biometrix, Reliance India Ltd, Emami, Society for Rural Initiatives for Promotion of Herbals (SRIPHL), Rural Community Assistance are actively working on this aspect.

Karnataka Watershed Development Agency and SUTRA are also involved in Jatropha plantation. Validated high yielding cultivars for different agro-climatic zones are yet to be identified. The ongoing S&T efforts of various institutes and universities on agro-technology of TBOs rest on the followings:

- Selection of superior germplasm for different agro-climatic zones
- Standardization of propagation techniques
- Standardization of agro-technology

Presently, genotypes collection, characterization , multiplication and field trials of Jatropha for cultivation in various states at different centers under NOVOD board and other organizations is going on with varying results. A centralized facility has been created at TATA Energy Resources Institute (TERI), New Delhi for oil characterization and at National Bureau of Plant Genetic Resources (NBPGR) for cryo-preservation of germplasm.

21.3.43. Intervention Needed in Jatropha Curcas

- Selection and development of suitable cultivars/varieties to improve the seed yield per plant and its oil content.
- Application of *in vitro* micro propagation for true to type and faster propagation of saplings.
- Utilization of Jatropha biomass for biogas production.
- Enhancement of oil content through Genetic Engineering.
- Region specific selection of Jatropha cultivars.
- Detoxification of Jatropha seed cake.
- Exploration of other uses of Jatropha leaves/fruits/barks/latex etc.

22. DELIVERABLES FROM JATROPHA

- High yielding Jatropha seed @ 3-4 MT/ha.
- Bio-diesel @ 1MT/ ha/ yr from 3rd year onwards.
- Protein rich cake @ 1.9 MT/ ha.
- Glycerine @ 0.095 MT/ha.
- Bio-fertilizers.
- Bio-pesticides.
- Bio-gas @ 500 m^3 / MT cake.
- Electricity - 2 MT cake digestion /day in methanation plant can run 25 KVA Genset for 8 hrs.
- Carbon credit @ $ 70 /ha/yr from 3rd year onwards.
- Training capsule to retiring troops for their better rehabilitation.

22.1 *Byproducts : Utilization*

The **glycerol, fatty acids salts and oil cake** are main byproducts from bio diesel production. However, various byproducts of *Jatropha curcas* are also of big use for human being.The research inputs are required towards utilization of glycerol for value added products and use of fatty acid salts as useful surfactants.

22.2 *Bio-Diesel*

Bio-diesel a more engine-friendly fuel is the main product for blending with petro-diesel as per norms and may be provided to ASC depot for its use in defence vehicles which will not only reduce import of diesel but also it will ensure reduced environmental pollution, rural economy and employment avenues. Quality strain yields seed @ 3-4 MT/ha and thus there will be estimated production of bio-diesel 1 MT/ha/yr from 3rd year onward.

22.3 *Protein Rich Cake*

Jatropha seed cake has been found to have a good feed value for animals because the nutritive potential of 18 species seed Kernel show crude protein 19:3 to 26%, lipid 43-59%, NDF 3.5-6.1 %, ash 3.4-5.0%, gross energy level 28.5-30.1 MJ/ Kg of meal (Makkar, et al. 1997). But the trypsin inhibiting activity is more. The seed cake although rich in protein is toxic to most of the animals which makes, it unfit for feed due to anti nutritional factor. A phorbal ester was found to be trypsin inhibiting and also the lectin and phytate do so (Martinez - Herrera, 2004). If cake is made free from these two factors it may be a valuable feed because the digestibility of protein is 86-90% almost higher than most of legumes. In one ha area there is estimated production of about 1.9 MT protein rich cake.

22.4 *Glycerine*

Glycerol is one of the important byproducts of the chemical transformation which has a great demand in the cosmetic and related industries like patent and pharmaceutical industry. Glycerine production is estimated @ 0.05 MT/ha and it may be marketed to pharmaceutical industry on competitive prices and thus revenue be obtained.

22.5 Bio-fertilizers

N:P:K in Jatropha oil cake is as good as in Neem oil cake, cow manure and poultry manure. By products, may be utilized within bio-fuel park for crop cultivation. However, the excess quantity may be disposed off in open market for revenue generation. N:P:K analysis of oil cakes of Jatropha, Non , Cow & Poultry manure is given in Table 21.

Table 21: Manure quality of Jatropha oil/cake.

S. No.	Plants	N	P	K
1.	Jatropha curcas	3.2-4.44	1.4-2.09	1.2-1.18
2.	Neem cake	4.85-5.0	0.36-1.00	1.03-1.5
3.	Castor cake	4.82	0.35	1.11
4.	Mahua cake	3.5	-	-
5.	Cow dung manure	0.97	0.69	1.66
6.	Poultry Manure	3.1	6.0	2.11

22.6 Bio-Pesticide

The insecticidal, fungicidal and toxic activity of *Jatropha curcas* have also been worked out. A remarkable effect of *Jatropha* against snail has been reported. The crude oil of *Jatropha curcas* was found to have contact toxicity to stored grain insects (*Sitophilus Sps.* and *Collosobruchus chinensis*) as reported by Solsoly (1996). In any experiments with Jatropha oil cake, tomato yield was obtained as good as treatment with Neem cake and carbofuran as compared to control (Table 22). The seeds are considered antihelminthic in Brazil and the leaves are used for fabricating houses against bed bugs. Plant extract of Jatropha has been reported as insecticidal and antifeedant against *Helicoverpa zea, H. virescence,Bemisia tabaci, Papilio demoleus* (Meshram et al.,1996 and Georges et al., 2008) and as mosquitocidal against *Aedes aegypti* and *Culex quinquefasciatus* (George et al.,2008 and Rehman et al.,2008). Haseeb,2007 reported deoiled cake of Jatropha as fungicidal against *Fusarium oxysporum,* nematicidal against *Meloidogyne incognita.* Zea and *H. Virescens,* antifedant against *Helicoverpa demoleus* (Meshram et al. 1996 and Georges et al., 2008) and as mosquitocidal against *Aedes agypti* and *Culex quinquefasciatus* (Georges et al., 2008) and Rehmam et al, 2008).

Table 22: Effect of Jatropha cake on root-knot (*Meloidogyne incognita*) nematode infecting tomato under field condition.

Treatment	RKI (0-5)*	Yield (kg/ha)
Jatropha cake @ 1.0 t/ha	3.54	27006
@ 2.0 t/ha	3.11	29815
@ 3.0 t/ha	2.84	32716
Neem Cake @ 2.0 t/ha	2.94	39568
Carbofuran @ 2.0 kg/ha	2.31	37037
Control	5.00	18271
S. Em. ±	0.24	1729
CV%	12.6	11.2

*0=Disease free; 5=Maximum disease intensity.

22.7 Bio-Gas

Bio-methanation co-produces bio-compost with biogas. Biocompost will go back to the soil to complete the bio-cycle and sustain the soil-humus high, to sustain yield and give freedom from chemical fertilizer. Bio-gas plant established in the vicinity of Military Farm will be utilized by the Military Farm to boil the water required for hygienic milking and also for cooking the food as a substitute to LPG with proper records. Biogas production is @ 500m^3/MT cake.

As per National mission on bio-diesel, Chhatishgarh State Govt 10 lakh fallow and degraded land has to be planted. It is estimated that by undertaking plantation of Jatropha/Karang on 10 lakh ha following production @ average 6 T seed/ha will accrue.

Jatropha seed	-	60 lac ton
Biodiesel	-	20 lac ton
Deoiled cake	-	40 lac ton
Glycerol	-	3 lac ton
Bio-gas	-	21000 lac
Electricity	-	400 MW

22.8 Biomethanation Technology

Most promising and immediately available source of clean energy from renewable sources is CH_4 which is also known by different names-

methane gas, gobar gas, natural gas, and producer gas. The best source of such energy is biomass- the substrate made of carbohydrates manufactured in the photosynthesis process in plants. In the anaerobic fermentation process, such as in the well-known gobar gas, bacteria decompose the substrate in a digester, without the presence of oxygen. The resultant gas is the biogas, mostly methane. The typical composition of the resultant gases will vary. Mostly it is about 60-70% CH_4, Carbon dioxide (CO_2) and some water vapour. If municipal wastes are used as a substrate, usually Hydrogen Sulphide gas H_2S is also released. Technologies are available for scrubbing the biogas to reduce the level of CO_2 and remove or reduce water vapour and Hydrogen Sulphide (H_2S). If biogas is used for energy, then it makes sense to improve its calorific value. The scrubbing of the biogas usually results in concentration of methane (CH_4) to about 80%. One of the options is to cultivate algae like *Spirulina* Blue green algae which is a rich source of beta carotene. The algae grow in a brackish water medium and the tank is closed at the top with a glass top to allow sunlight. The bio gas is bubbled up through the tank and CO_2 gets absorbed in the algae photosynthesis, resulting in an improved calorific value with higher concentration of CH_4. Hence the useful byproduct *Spirulina* can be cultivated on a large scale, which can help bridge the per capita protein deficiency in India. This cultivation improves photosynthesis due to CO_2 absorption and hence algae yield is better for same unit area. (Anonymous 2005).

Removal of H_2S is necessary because when burned in an engine it causes formation of sulphur dioxide and ultimately sulphuric acid, which corrodes and reduces the life of the engines. Maybe this sulphur can be added back into the last stage of vermi compost since sulphur is a deficient micro nutrient item in many agricultural systems, after world-wide preference to low suphur crude leading in turn to less "acid rain" (http://www.vkrao.com/biotech.htm).

22.9 *Use of Methane Gas*

- Cooking and heating in cities and is normally transported by a pipe system.
- Generating power or electricity through micro turbine, diesel/gas/dual fuel engines. However, it is advisable to scrub the gas to remove H_2S and NH_3 when it is used in any type of internal combustion engine. When the gas is clean, it can be used in

transportation also such as city bus and automobiles, however, compression is the only problem.

- The residual mass left over can be used as manure. This manure derived out of Methane preparation which is also called as biogas manure, is very useful for soil/crops. The solid residuals can also be treated with bio-sanitizer for better stabilization of pathogens and also retard mineralization.

The benefits of methane gas as renewable energy source are enormous. It will cut down on the foreign exchange drain on India's import of crude oil, which will be largely replaced by CH_4 produced from renewable sources. It will reduce the emission of Carbon, since such wastes release methane gas anyway by natural aerobic fermentation, but is now made available in a controlled process and its energy is used and it replaces equivalent amount of fossil fuels which would have otherwise been burnt. This means a reduction in levels of carbon emissions as Green House Gases into the atmosphere. In today's global warming threat looming large over the world, this is an important step in bringing down carbon emissions from India.

In the context of global warming, CO_2 emissions are projected to grow from 5.8 billion tonnes carbon equivalent in 1990 to 7.8 billion tonnes in 2010 and 9.8 billion tonnes by 2020. Much of it is expected to come from developing world- 81% of the increase during 1990-2010 and 76% of the increase between 1990- and 2020. The developed world of course would not like to give up their energy sources- mostly from fossil fuels. They would like to bury CO_2 into the Earth! (Australia)

Small farmer with a few cattle and large farms, herds and cattle sheds can resort to methanation. Villages can go in for community digesters for human excreta steams. The available gases can be used for lighting, cooking needs of the rural inhabitants. Women will experience less drudgery for daily fuel needs. Dr. A.D. Karve's digester uses carbohydrates like spoilt food and damaged grains to generate comparative larger volumes of methane as compared to digesters using excreta as the substrate. This innovation is a welcome development as it enables wastes to be put to good use.

22.10 Electricity

2 MT cake digestion /day in methanation plant can run 25 KVA Genset for 8 hrs. The electricity thus generated, will be utilized for

various activities like irrigation of planted crop, decortications & dehulling, oil extraction and trans-esterification and also for routine activities.

22.11 Carbon Credit

Chaturvedi, (2006) reported that the proposed large scale Jatropha plantation in India will substantially reduce green house gases in the atmoshphere by absorbing CO_2 from atmosphere. This will also provide opportunity to the State for trading Co_2 internationally, to the countries having much higher pollution level, in the light of Kyoto Protocol. **On an average 1 ha of Jatropha plantation absorbs 10 tonnes of CO_2 from the atmosphere and as such 1 million ha shall result in absorption of 10 MT of CO_2 from atmosphere.** Even on the basis of prevailing trading rate of $ 7.00 per tonnes of CO_2, State will be in a position to trade CO_2 to the extent of more than Rs. 300 crore, internationally every year.

Carbon credit of one MT of bio-diesel leads to reduction of 3 MT of CO_2 emission. Carbon credit possibilities will be explored under Kyoto Protocol and thus foreign exchange will be generated for the nation. Carbon credit with one ha of Jatropha cultivation is given below:

22.11.1 Carbon Trading

The overall goal of an emissions trading plan is to reduce emission. The cap is usually lowered over time-aiming towards a national emissions reduction target. In other systems a portion of all traded credits must be retired, causing a net reduction in emissions each time a trade occurs. In many cap and trade systems, organizations which do not pollute may also participate, thus environmental groups can purchase and retire allowances or credits and hence drive up the price of the remainder according to the law of demand. Corporations can also prematurely retire allowances by donating them to a nonprofit entity and then be eligible for a tax deduction.

Because emissions trading uses markets to determine how to deal with the problem of pollution, it is often touted as an example of effective free market environmentalism. While the cap is usually set by a political process, individual companies are free to choose how or if they will reduce their emissions. In theory, firms will choose the least-cost way to comply with the pollution regulation, creating incentives that reduce the cost of achieving a pollution reduction goal. Emissions trading principles are based on proposals by the Technocracy movement of the 1930's. Technocracy proposed a system of Energy Accounting, or emission trading, to promote balanced and harmonious development throughout the world.

22.11.2 Cap & Trade

The textbook emissions trading program can be called a "Cap & Trade" approach in which an aggregate cap on all sources is established and these sources are then allowed to trade amongst themselves to determine which sources actually emit the total pollution load. An alternative approach with important differences is a baseline & credit program. In a baseline and credit program a set of polluters that are not under an aggregate can create credits by reducing their emissions below a baseline level of emissions. These credits can be purchased by polluters that are under a regulatory limit. Many of the criticisms of trading in general are targeted at baseline and credit programmes rather than cap type programmes.

22.11.3 Economics of international emissions trading

It may be possible for a country to reduce its emissions using a Command-Control approach with regulation, direct and indirect taxes. The problem with such an approach to abatement is that it may cost the economy more to reduce the same amount of pollution when compared to the 'Emissions Trading' scenario. The economic reason behind this extra cost is because of the different Marginal Abatement Costs (MAC) for taking action in different countries. The Marginal Abatement Cost refers to the cost spent to reduce an extra unit of pollutant or other emissions. Taking advantage of the difference in MAC's is the principle behind the international emissions trading markets.

22.11.4 Kyoto Protocol

The Kyoto-Protocol is a 1997 international treaty which came into force in 2005, which binds most developed nation to a cap and trade system for the six major green houses. Emission quotas were agreed by each participating country, with the intention of reducing their overall emissions to 1990 levels by the end of 2012. Under the treaty, for the 5 year compliance period from 2008 untill 20.

12, nations that emit less than their quota will be able to sell emissions credits to nations that exceed their quota.

It is also possible for developed countries within the trading scheme to sponsor carbon projects that provide a reduction in greenhouse gas emissions in other countries, as a way of generating tradable carbon credits. The protocol allows this through Clean Development Mechanism (CDM) and Joint Implementation (JI) projects, in order to provide flexible mechanism to aid regulated entities in meeting their compliance with their caps. The intergovernmental Panel on Climate Change has projected that the financial effect of compliance through trading within the Kyoto commitment period will be 'Limited' at between 0.1-1.1% of GDP among trading countries. This compares with an estimate in the Stern report which placed the costs of doing nothing at five to 20 times higher.

One of the most important CDMs that are emerging is the system of carbon trading, which allows the development of a market wherein carbon dioxide as well as carbon equivalents, i.e, other greenhouse gases like methane, can be traded between participants. The participants could be countries or companies. Though the political and institutional framework for carbon trading is yet to develop, it is generally believed that a potentially large and lucrative global market for carbon trading could develop by the end of the decade. (Dadwal, 2003, The Financial Express).How does the system work? Once the KP enters into force, developed countries are required to reduce their average Green House Gases emissions by 5% by 2008-12. A country or company wishing to reduce or meet their emission targets can do so by investing in clean projects, which would contribute towards offsetting their GHG emissions, but would also earn the investor some "credits" which would go towards a net carbon reduction. A typical CDM project would be substituting fossil fuel-based power generation with renewable energy or a project that would improve existing energy efficiency levels. Or, as in India, by investing in forestation or community tree planting projects, called "carbon sinks".

Currently, carbon trading projects take place within some countries including the US and UK, though recently some trade has also taken place between countries, as well. But the potential for inter-state trade has been estimated at around $2 trillion over the next 10 years. However, for a full-fledged carbon trading market to develop, it would be necessary for the KP to come into force as, according to experts, trading would only make sense if companies operated under emissions caps set by their governments. Without an overarching regulatory mechanism, the system would at best operate informally, providing no incentive for emissions reductions. That is why Russia's accession is deemed crucial. However, according to some environmental experts, even if Moscow decides against coming aboard the KP, there is a way out. As per a clause in Article 20 of the protocol, an amendment to the treaty could be adopted by the parties to the Protocol, preferably by consensus, but if not, as a last resort by three-fourths majority vote of the parties, whereby the goal of 55% reduction in emission levels could be reduced and hence allow the treaty to come into force, which would, in turn, take the pressure off the advocates of the treaty to get the requisite countries aboard.

22.11.5 Greenhouse effect : global warming

CO_2 molecules in the air absorb the heat radiated from earth's surface though it does not absorb the energy coming from the sun. The dual behavior of CO_2 causes the green house effect. As the use of fossil fuel is increasing and forests which absorb CO_2 in the process of photosynthesis are reducing, the concentration of CO_2 is increasing in the atmosphere. With the increase in the concentration of CO_2 the earth is getting warmer and warmer. It is feared that by the year 2100 AD global temperature will increase by 15^0 C. This continuous rise of the concentration of CO2 may result in change in rain fall pattern and melting of ice which may engulf many costal areas. 5% rise in global surface temperature is expected by the end of the century and 25 billion tonnes of C emissions need to be cut to contain warming. 55% of World's carbon emission is produced by 15% of population (Chengappa 2008).

Thus bio-diesel production through non-edible oils involves plantation of *Jatropha curcas / Pongamia* in huge areas which will help in absorption of CO_2 in the process of photosynthesis and finally cleaning of the atmosphere.

22.12 Medicinal Use

In addition to an important biofuel, *Jatropha* has been explored for many medicinal properties also. An anti-tumour crucin compound isolated from *Jatropha* was found to have anti-tumour effect due to N-glycoside activity. Jatropha contains an alkaloid known as Jatrophine - one, a novel monocylic diterpine was found to have strong antibacterial activity (Ravindranath et al., 2003). The latex of *Jatropha curcas* was found to reduce blood clotting time at higher concentration. Three deoxypreussomerins, palmarumycins CPI, JC 1 and JC 2 from the stem of *Jatropha* exhibited strong antibacterial activity against/ *Staphylococus aureus,* which was comparable to that of standard compound penicillin - G, Anti-inflammatory activity of *Jatropha* roots in topical application of methanol extract exhibited systemic and significant anti-inflammatory activity in acute carrageenan induced edema in rats (Majumdar et al., 2004). Latex of *Jatropha curcas* traditionally used as wound healer and anti wart activity was explored for antibacterial activity against five skin infections viz., *Enterococcus faccialus, Escherichia coli, Pseudomonas aeuginosa, Streptophylococcus aureus* and *Staphylococcus epidermidis.* It was observed that the fresh latex inhibit the growth of all bacterial species. Latex has anti-cancerous properties. It can heal wounds and also has anti-microbial properties. It is also used as an external application for skin diseases and rheumatism and for source on domestic live stock.

Anti-metastatic effects of curcusone - B from *Jatropha curcas,* a diterpene was investigated in 4 human cancer cell lines. The non cytotoxic dose of curcusone - B resulted in a strong reduction of *in vitro* invasion, mortality and secretion of matrix metallo proteins (MMP) of the cancer cells. It was thus concluded that curcusone - B thus effectively supresses the metastasis process.

The fruits, stems, leaves, roots and latex as described earlier has a wide range of uses in traditional medicine, therapies and cosmetics and also for the treatment of many diseases of modern originThe denudates of the plant are used for cleaning teeth and juice of the leaf is used as an external application for piles. The roots are used as an anti-dotes for snake bite. The Jatropha oil is strong purgative and widely used as an anti-septic for cough. Jatropha seeds are used against constipation; the white latex serves as a disinfectant of mouth infections of small children. It also stops bleeding. The leaves are used against malaria and for massage of luxations (http://www.jatropha/faq,htm.).

22.13 Dye, Soap and Illuminant

The bark of Jatropha plant yields dark blue dye which is used for colouring clothes, fishing nets and lines. Oil has a very high saponification value and is being used for making soap in some countries. Also the oil is used as an illuminant as it burns without emitting smoke.

22.14 Training and Employments

Training will be imparted to service personnel free of cost for their acquaintance on bio-diesel production and for undertaking such work at small scale for their better rehabilitation.30 million hectares of Jatropha cultivation for bio-diesel can completely replace the current usage of fuel in India,. Use of 11 million hectares of waste land for Jatropha cultivation can lead to generation of minimum 12 million Jobs.

23. OIL EXTRACTION PROPERTIES OF JATROPHA

An overview of bio-diesel (fatty acid alky esters), the potential fraction of *Jatropha curcas* has been worked out by Kantwal and Soni, 2003. Fatty acid composition and properties of *Jatropha* seed oil and its methyl ester has also been studied by Chatakananda et al. (2005). These authors observed that seed contained major component as lipid (46.3%). The oil fraction contained 78.4 % unsaturated fatty acid (Oleic acid 42.2%), Linoleic acid (35.2%) and 21.7% saturated fatty acid in which Palmitic acid was 14.7% and stearic acid 6.9 %. The chemical properties, acid value, peroxide value, iodine value, saponification value and trans-esterification by methanol and sodium hydroxide were studied (Table 23). The yield of methyl ester was found to be 80.45 % (Table 24). The study concluded that methyl ester met the standard for bio-diesel. The importance of *Jatropha* bio-fuel for socio-economic development and cleaner environment particularly the fatty acid alkyl esters have been described by Kantwal and Soni, (2003).

Table 23: Oil of *Jatropha*

Saponification value	23.13
Iodine value	103.92
Acid value	1.41
Peroxide value	12.67
Unsaponifide matter	1.50
Free fatty acid	2.80
Sp. Gravity	0.92

Table 24: Proximate composition of *Jatropha curcas* seed oil.

Parameter	Amount
Lipid as major component	46.3%
Oil extraction efficiency	
(i) Haxane	94.6 % 3 hrs at room temperature
Oil fraction unsaturated Fats	78.4%
(i) Oleic acid	42.2%
(ii) Linoleic acid	35.2%
Saturated fats	21.7%
Palmitic acid	14.7%
Stearic acid	6.9%
Methyle ester yield	
(MeOH +NaOH as catalyst)	80.45%
Meet all the standard of Bio-diesel	

An experimental comparison of methanol and *Jatropha* oil in a compression ignition engine was done by Kumar et al. (2003). In various methods of using vegetable oil (*Jatropha* oil) and methods such as blending, trans-esterification and dual fuel operation were studied experimentally. A single cylinder direct injection diesel engine was used for this work. Tests were done at constant speed of 1500 rpm, min-1 at varying power outputs. In dual fuel operation the methanol to Jatropha oil ratio was maintained at 3:7 v/v and this was close to the fraction of method used to prepare the methyl ester. Result indicated that Brake thermal efficiency was better as **Dual Fuel Operation** (DFO) and with methyl ester of *Jatropha* oil as compared to the blend. The increase was from 27.4% with *Jatropha* oil to maximum 29% with methyl ester and 28.7% in dual fuel operation. Smoke was reduced in all methods compared to neat vegetable oil operation. Value of smoke emission was Bosch Smoke Unit (BSU) with neat *Jatropha* oil - 4.4, BSU with the blend was 4.1, with methyl ester was 4 and 3-5 with dual fuel operation. The nitric oxide (NO) level was lower in *Jatropha* oil compared to diesel which further reduced in DFO and blend with methanol. But dual fuel operation showed higher Hydrocarbon and CO emission than ester and the blend.

Extraction of oil from *Jatropha curcas* L. seed Kernel by combination of sonication and aquous enzymatic oil extraction was done by Shah et al. (2005) and these authors reported that use of ultrasonication as treatment before aquous oil extraction gave an yield of 67% (10 min ultrasonication as pH 9 followed by aquous oil extraction). However, the maximum yield of 74% was obtained by ultrasonication for 5 minutes followed by aqueous enzymatic oil extraction using an alkaline protease at pH 9. Various methods of oil extracting of *Jatropha* have been standardized. Winkler (1997) reported that the hexane extracts 98% of oil and water only 38% of total oil seeds. Use of several cell wall degrading enzymes during agencies extraction the oil yield goes up to 86% when alkyl protease was used. Use of ultrasonication also resulted in reducing the process time from 18 hrs to 6 hrs.

Eighteen different provenances of *Jatropha curcas* from countries in West and East Africa, North Central America and Asia were characterized by Makkar et al. 1996 & 1997 for nutritive potential and toxic constituents (anti-nutritional factor). Mean weight of seed of 18 provinces was 0.64 - 0.76 g. Kernel forms large proportion of seed and accounts for 61.3 - 64.5%, crude-protein varied 19.3-26.0%. Lipid (43.59 - 53.0%), Neutral detergent fibre (3.5 - 6.1 %) and ash (3.4 - 5%) in Kernels. The gross energy level of Kernel defatted, was (28.5 - 31.2 MJ/ kg). Trypsin inhibitor activity in defatted meal varied from 18.4 - 27.5 mg of trypsin inhibited/g. Saponins varied 1.8 - 3.4% as diosgenin equivalent and Phytic acid 6.2 - 10.1 %, however, amylase inhibitor was not found.

23.1 Plant Lipids

Analysis of Jatropha seed shows the chemical composition viz., moisture (6.20%), protein (18.0%), fat (38.0%), carbohydrate (17.0%) fiber (15.50%) and ash (5.30%). The oil content varies 35-40% in the seed and 50-60% in the kernel depending upon the storage period of the seed. The oil contents 21% saturated fatty acids and 79% unsaturated fatty acids. Lipids are the major form of carbon storage in the seeds of many plant species, constituting up to 80% of the dry weight of such seeds and vegetative cells of plants contain 5 to 10% lipid by dry weight, and almost all of this weight is found in the membranes (Ohlrogge and Browse, 1995). In most plants, storage lipids are in the form of tri-acyl-glycerol i.e. TAG (Murphy, 1999). In contrast to this, Jojoba, a desert plant found in American Southwest, is the only plant species known to accumulate waxes up to 60% dry weight as storage lipid (Thelen and Ohlrogge, 2002). Seed storage lipids are an energy store and food reserve for germination and post germinative growth of seedlings and a repository of essential and non- essential fatty acids for phospholipid biosynthesis (Tzen *et al.*, 1993).

Storage lipids may be accumulated in one or both of the main types of seed tissue, embryo or endosperm. In oilseeds such as sunflower, linseed or rapeseed, the cotyledons of the embryo are the major sites of lipid accumulation. In species such as castor bean, coriander or carrot, the endosperm is the main site of lipid accumulation and in tobacco, both embryo and endosperm tissues store lipids (http://www.cyberlipid.org). Vegetable oils are classified into two main groups, according to their source: pulp oil (Palm, Olive, Avocado) and seed oil (other sources). The amount of lipids in plant parts varies from as low as 0.1% in potatoes to about 70% in Pecan nuts. Some vegetable products are fat poor (1% in lentils) and some seeds have a middle range amount (about 10% in wheat germ).

More than 300 different fatty acids are known to occur in seed TAG (Harwood, 1980 and Van de Loo *et al.*, 1993). Their structures may vary in chain length from as few as eight carbons to >22 carbons. But just five fatty acids account for 90% of the commercial vegetable oil produced (Gunstone,1986 and Hilditch and Williams, 1964). These fatty acids are most commonly found in membrane lipids. In addition, certain plant species have the capacity to produce specialized fatty acids that are used for TAG synthesis but that are largely excluded from the membrane lipids. The position and number of double bonds may be unusual, and various functional groups such as hydroxy or epoxy may be added to the acyl chain.

The current volume of traded vegetable oil is over 70 MT per year and is predicted to be over 100 MT per year by the year 2010 (Murphy, 1996). The four most important oil seed crops are, in descending order i.e. Soybean, Oil palm, Rape seed and Sunflower, which together account for 65 % of current world wide vegetable oil production (Gunstone, 2001). The fatty acid profiles of these oils are reported by Ghotra *et al.*, (2002). The vast majority of vegetable oils is currently used for edible commodities such as cooking oil, margarines and processed food. Only about 15% goes towards the manufacture of oleo chemicals i.e industrial products derived from oil crops such as soaps, detergents, lubricants, bio-fuels, cosmetics and paints. Fatty acid composition of selected plants and bio-chemcial composition of *Jatropha curcas* and its related species are given in Table 25 &26.

Table 25. Fatty acid composition (%) of selected plant oils

Fatty acids	Linseed oil	Sunflower oil	*Jatropha curcas* oil	Rice bran oil
Palmitic acid (16:0)	7.35	4.65	13.05	19.66
Oleic acid (18:2)	38.27	61.07	45.59	50.88
Linolenic acid (18:1)	8.85	23.83	28.86	27.21
Linoleic acid (18:3)	44.41	1.49	5.99	1.37
Archidic acid (20:0)	0.99	1.44	1.12	0.71
Bahenic acid (22:0)	0.13	7.53	5.39	0.17

Table 26. Bio-chemical composition of *Jatropha curcas* and related strains

Parameter	*J. curcas*	*J. glandulifera*	*J. gossypifolia*	*J. multifida*
Total oil (%)	30-40	23-30	28-34	30-35
Myristic acid(%)	0-1.6	0.4-2.3	1.3-20.9	0-0.8
Palmitic acid (%)	9-21	12-14	10-30	15-24
Steraric acid	2-10	6-8	0-3.5	3-6
Arachidic acid (%)	0.1-2	0	0-0.7	0-0.8
Arachaedic acid (%)	35-64	23-34	10-38	18-3
Oleic acid (%)	18-45	43-51	41-74	42-5
Linoleic acid (%)	16-31	21-23	15-24	25-27
Total unsaturated fatty acid	69-79	60-69	26-65	50-70

23.2 Trans-Esterification-Definition and Properties

Trans-esterification is the most common method and leads to monoalkyl esters of vegetable oils and fats, now called bio-diesel when used for fuel purpose. Basically it consists of reacting the vegetable oils feedstock with an alcohol, usually methanol, in the presence of a catalyst, usually a base viz., sodium or potassium hydroxide, to give the corresponding vegetable oil (usually methyl esters). Methyl esters are the most common form of bio-diesel, largely due to methanol being the least expensive alcohol. Bio-diesel is made through as chemical process called trans-esterification whereby the Glycerine is separated from the fat are vegetable oil. The process leave behind two products-

(1) Methyle ester (chemical name for bio-diesel) and (2) Glycerine (a valuable byproduct to be used in soap and other products).

Chemically all vegetable oils, whether edible or non-edible, and fats are made up of fatty acid tri-glycerides. This means that even though there is change in the sources of edible oils, chemically they remain almost same. These tri-glycerides when reacted with an alcohol (usually methanol or ethanol) in the presence of an acid or a base catalyst very readily give quantitative amounts of fatty acid esters. This reaction is called **trans-esterification** and the product obtained i.e. fatty acid esters is called **bio-diesel (Tuli, 2006)**.

In trans-esterification, KOH and methanol are mixed to create potassium methoxide (K^+ $CH3O^-$). When mixed in with the oil this strong polar-bonded chemical breaks the transfatty acid into glycerine and ester chains (bio-diesel), along with some soap if you are not careful. The esters become methyl esters. They would be ethyl esters if reacted with ethanol instead of methanol. (http:// www. vkrao.com/ biotech.htm).

Trans -esterification reaction was developed almost a century ago and has been used in chemical industry very frequently. The chemical kinetics and the reaction variables have been thoroughly examined. Depending upon the catalyst used for trans-esterification/ esterification, commercial bio-diesel technologies can be divided into three categories.

- Base catalysed trans-esterification with refined oils.
- Base catalyst esterification with low fatty acid oils and fats.
- Acid esterification followed by trans-esterification of low or high FFA (Free Fatty Acid) fats and oils.

Trans esterification also called alcoholysis, is the displacement of alcohol from an ester by another hydrolysis alcohol in a process similar to hydrolysis. Methanol is most commonly used for the purpose since it is the cheapest alcohol available. Ethanol and higher alcohols such as isopropanol, butanol etc can also be used for the esterification. Using higher molecular weight alcohols improves the cold flow properties of bio-diesel but reduces the efficiency of Trans-esterification process. The reaction is given below:

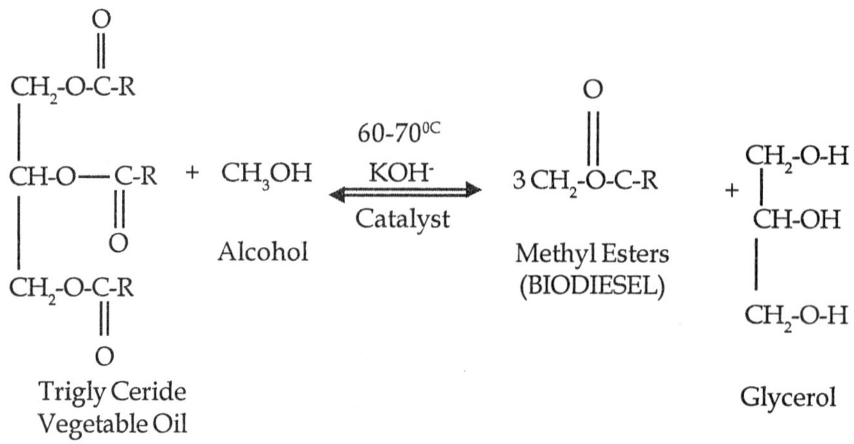

Chemistry of Trans-Esterification

When FFA content of the oil is above 1%, difficulties arise due to the formation of soap, which promotes emulsification during the water washing stage. If the FFA content is above 2%, the process becomes unworkable. The factors affecting the trans esterification process are (i) Oil temp (ii) Reaction temp. (iii) Ratio of alcohol to oil (iv) Catalyst type and conc. (v) Intensity of mixing (vi) Purity of reactant.

The currently available processes for production of bio-diesel include acid catalyzed trans-esterification, base catalyzed trans-esterification and conversion of oil-glycerides to fatty acids and then esterification of the produced fatty acids to bio-diesel (Fig 52, 53 & 54). The existing processes have limitations in terms of efficiency, process ability of oils of high viscosity, high free fatty acid content and high-saturated oil-glycerides. The conversion yield, separation of unwanted reaction products and purification of the produced bio-diesel are major problems being studied for their effective and reliable solution. Work related to use of enzymes for bio-diesel production is being carried out in certain Indian Laboratories and an innovative catalyst-free process so developed shall provide an edge to India in bio-diesel production technology.

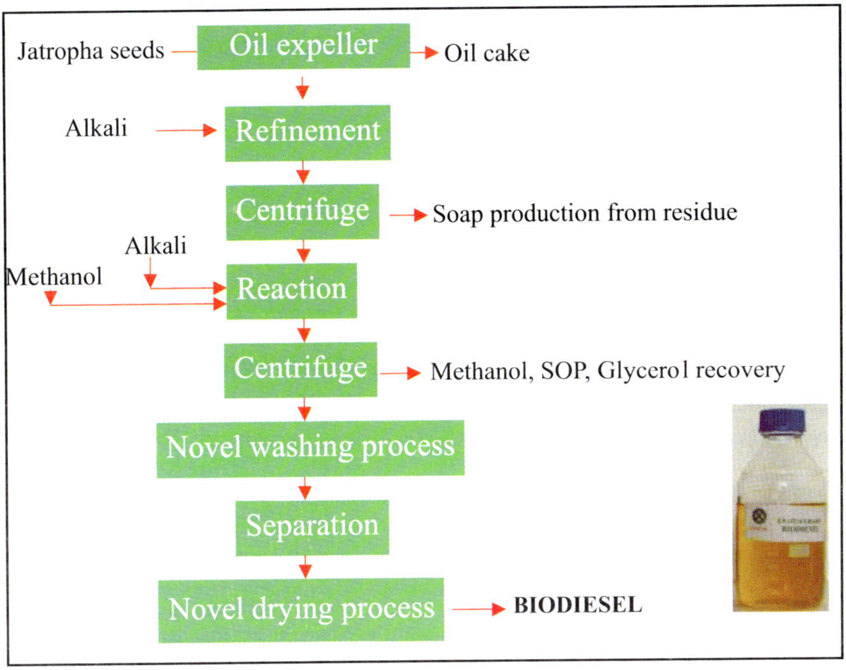

Fig.52. Schematic block diagram for the process for Bio-diesel from Jatropha seeds

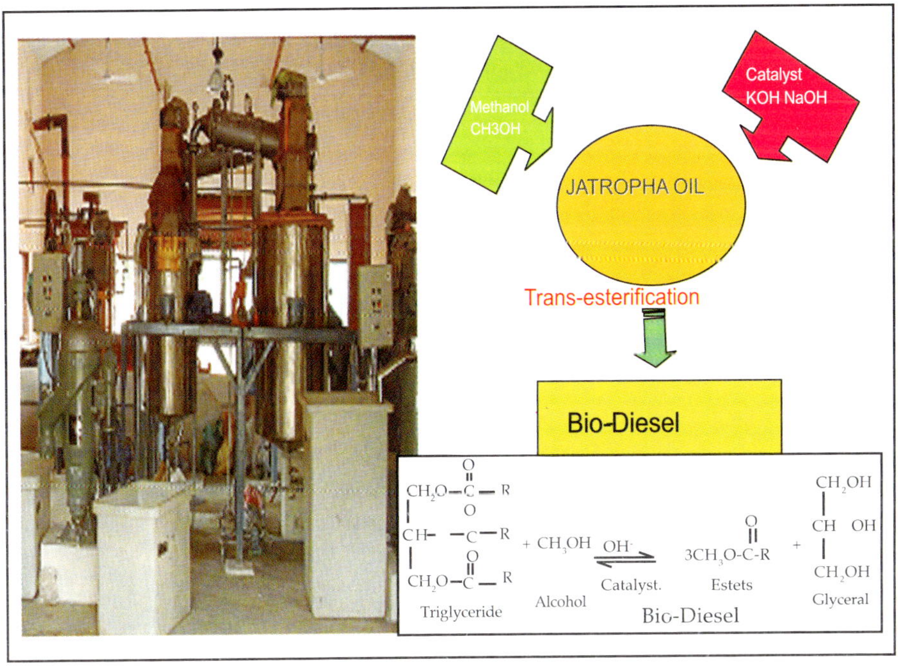

Fig. 53. Processing : Bio-diesel

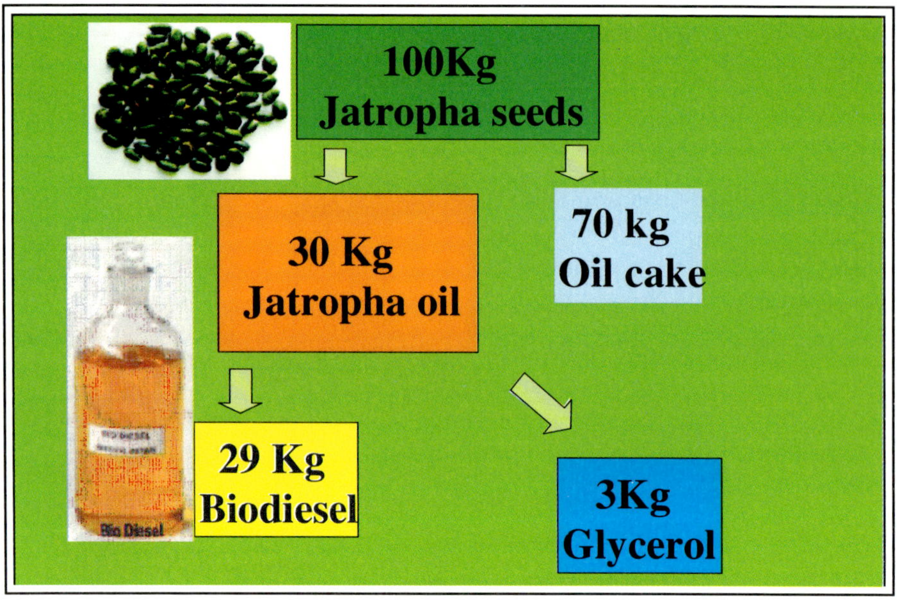

Fig. 54. Bio-diesel from Jatropha seed

Presently, alkali catalyzed trans-esterification of vegetable oil and fats is most commonly used method to produce bio-diesel. However, this method is not of choice for high FFA feedstock and has several disadvantages, such as removal of catalyst, deactivation of catalyst by moisture, by-product formation, waste water treatment and multiple steps to purify bio-diesel to meet the specifications. To overcome these problems, newer methods of trans-esterification using bio-catalyst, super critical alcohol and hetrogenous catalyst are being explored.

There have been some reports of laboratory scale experiments using heterogeneous catalyst which catalyses trans-esterification and esterification simultaneously (IIP,Dehradun), solid fuel catalyst called double metal catalysts for continuous trans-esterification process (IICT Hyderabad and NCL, Pune) but these are yet to achieve commercial maturity. IICT, Hyderabad developed a continuous transe-sterification process @ 1L/hr which is under upscaling to 10 L/ hr. This institute is also studying lipase mediated transformation of vegetable oil into bio-diesel using propan 2 ol as acyl accepter (Modi et al., 2006) Looking at the commercial maturity, the work regarding trans-esterification as such needs to be strengthened to upscale the existing technology and to establish techno-economic viability. Central Saline Marine & Chemical Research Institute (CSMCRI), Bhavnagar has developed trans-esterification plant with 1 tonne per day (1TPD) capacity to produce bio-diesel conforming to ASTM D 6751 and EN 14214 standards. Delhi College of Engineering has developed 600 L per day capacity plant, which is under upscaling to 1TPD. Eastern and Southern Railways have installed 1TPD capacity plant at Kharagpur and Perambur (Chennai) in collaboration with Indian Oil Corporation. DRDO is collaborating with leading Indian Institutes i.e. IIT Guwahati, IIP Dehradun, IICT Hyderabad, CSMCRI Bhavnagar, and will have a plan to collaborate with DCE, Delhi and industries- Mint Bio-fuel Pune and IOC, Faridabad in respect of bio-diesel production with special reference to their technological innovations and capabilities.

Despite all the development work at national level, there are few technology gaps which need to be looked into at the present :

- Soap formation leading to high usage of water.
- High FFA which interferes with trans-esterification process with lower bio-diesel yield.
- Problems of effluents.
- Minimum consumption of energy.
- Cost of processing so that it can be adopted as small scale industry.
- Upscaling of trans-esterification process without compromising on bio-diesel quality.

23.3 Institutes Engaged on Trans-Esterification Process in India

- Central Saline and Marine Chemical Research Institute, Bhavnagar, Gujarat
- Indian Institute of Chemical Technology, Hyderabad
- Indian Institute of Petroleum, Dehradun
- National Chemical Laboratory, Pune
- Indian Institute of Technology, Guwahati
- Indian Institute of Sciences, Bangalore
- Tamil Nadu Agriculture University, Coimbatore
- Delhi College of Engineering, Delhi
- Punjab Agricultural University, Ludhiana
- Indian Oil Corporation (R&D), Faridabad
- Indian Institute of Technology, Kanpur

- Indian Institute of science
- Railway workshop
- NOVOD Board
- Chhittishgarh Biodiesel Development Authority
- Defence Agricultural Research Laboratory (Uttarakhand).

23.4 Storage of Bio-Diesel

The storage studies on plant oils and their methyl ester has been done by Sanghager et al. (2004) and reported that esters of plant oils are preferred to parent oils if used as fuels in diesel engines because the ester fiction reduces the viscosity of plant oils and makes it comparable to that of diesel. The storage studies of oil and its esters with respect to free fatty acid contents, viscosity and change in weight for one and half year were done on linseed oil ester, sunflower oil ester and rice bran oil ester. The maximum FFA content % change was in linseed oil followed by sunflower and rice bran ester where as *Jatropha* oil showed minimum change. The increase in FFA contents was below to BIS requirement. The viscosity level in all the oil ester was similar initially but after six months of storage the viscosity increased by 1.6 times in linseed oil ester and sunflower ester where as bran oil showed 1.4 and *Jatropha* oil only 1.3 time increase in viscosity. All other oil esters showed increased weight indicating absorption of moisture or reaction with oxygen in air, however, *Jatropha* seed oil showed negligible increase in weight. The effect of these oils on engine parts was also studied during contact period of two months. No significant change in weight of metallic components was observed where as in non metallic components, such as rubber parts significant changes were observed. (Abigor, et al, 1997). The seed contained 50.50% oil, 14.78% protein, 4.74 % moisture, 3.33 % carbohydrate, 12.25 % ash and 2.5% fibre. The saponification value was 23.13, Iodine value was 103.92, Acid value was 1.41, Peroxide Value was 12.67 unsaponifiable matter was, 1.50 free fatty acids were 2.87 and specific gravity was 0.94. The fatty acid composition was 0.35% , Dodecanoic acid, 0.47%, Myrristic, 27.04%, Palmitic Steric 3.15% and 69.00% Oleic acid of total fatty acids. The Kernel: Shell ratio was found between 61:39 to 64:36 and oil contents from 32.4 to 39.0%. Oil was found rich in oleic and linoleic acids and contained 11.4 - 15.5 % plamitic acid. Iodine value varied from 97 to 107 indicating needs hydrogenation.

Table 27: Physicochemical properties of seed and seed oil of *Jatropha curcas* (Abigor, R. A. et al., 1997 Nigeria).

Parameter/Component	Amount	Parameter	Amount
(A) seed oil	50.50%	Protein	14.78%
Moisture	4.74%	Carbohydrate	3.33%
Ash	12.25%	Fiber	3.50%
(B) Oil saponification value	23.13	Acid value	1.41
Peroxide value	12.67	Unsaponifiable	1.50
Free fatty acid	2.87	Matter	-
Iodine value	103.42	Specific gravity	-
(C) FFA in oil Dodecaroic acid	0.35%	Myristic acid	0.47
Palmitic acid	27.05%	Stearic acid	3.15
Oleic acid	69.00%		

Long term storage of bio-diesel leads to increase in FFA and viscosity due to auto-oxidation and polymerization. Adequate data on long-term storage of bio-diesel and blends are not available. Based on the experience of producers, bio-diesel can be stored up to a maximum of 6 months period. The existing storage facilities and infrastructure for petrol and diesel can be used for the bio-diesel with minor alteration. For bio-diesel storage, study of shelf life and how it might break down under extreme conditions assume importance.

Since bio-diesel is made using biological based products, the temperature at which it is stored is more critical than in the case of petroleum diesel. It can grow moulds if stored at high temperature. On the contrary, it will thicken and could become difficult to dispense if stored at very low temperature. The viscosity variation among different esters could be attributed to the variations in the fatty acid composition of the esters. More the unsaturation, more is the polymerization. In a storage study carried out at Punjab Agriculture University, Ludhiana, the change in FFA content, viscosity and weight, after one year of storage was found to occur. However, it was minimum in *Jatropha curcas* oil ester as compared to linseed, sunflower and rice bran oil esters (Singh et al., 2004).

The efficient storage of bio-diesel resources can provide energy security to the country. Adequate data are not available for long-term storage of bio-diesel and blends. Based on the experience, bio-diesel can be stored up to a max. 6 months. As a mild solvent, bio-diesel tends to dissolve sediments normally encountered in old diesel storage tanks. Brass, Teflon, Lead, Tin, Copper, Zinc etc. oxidize bio-diesel and create sediments. The existing storage facilities and infrastructure for petrol and diesel can be used for the bio-diesel with minor alterations. Most of the advanced countries have finalized their bio diesel specification.

Salient properties of bio-diesel are given below:

- Bio-diesel has poor oxidation ability. Use of oxidation stability additives is necessary to overcome this problem.
- Low temperature can cause bio-diesel to gel, but on warning it liquefies quickly. Hence, insulation/jacketing of storage tanks and pipelines would need to be done at the low temperature zone.
- To avoid oxidation and sedimentation of tanks with bio-diesel, storage tanks made of aluminum, steel etc may be tried.

Indian Institute of Technology, Delhi has studied the effect of two antioxidants-TBHQ (Tert Butyl Hydro Quinone) and BHA (Butylated Hydroxyl Anisone) on shelf life of biodiesel at 30 ^{0}C. TBHQ was found to enhance the shelf life from 18 days (100 ppm) to 140 days (1000 ppm), whereas BHA @ 1000 ppm maintained bio-diesel quality up to 196 days at 20 ^{0}C and 98 days at 30 ^{0}C. High altitude storage studies are going on in one of the DRDO Lab i.e. Defence Institute of High Altitude Research, Leh.

Automobile industries like Daimler - Chrysler, India have set up storage and dispensing facility for bio-diesel at Pune to facilitate fleet-testing and research upon shelf life and storage issues of biodiesel. However, no systematic studies have so far been conducted and this needs immediate attention to sustain bio-diesel programme. It becomes quite important from defence point of view specially at high altitudes, in deserts and high humid areas.

24. GENETIC ENGINEERING

To develop quality planting material for mass production, different laboratories and institution viz., State Forest Research Institute (SFRI), Jabalpur, GBPUA&T, Pantnagar, Regional Research Laboratory (RRL), Bhabneshwar, Chaudhary Charan Singh University, Meerut, Hissar Agriculture University, Karnal , IIT, Gwahati, CSK HPKV, Palampur and different intuitions are working on mass propagation of Jatropha.

Institutes like IIT,Guwahati; TNAU,Coimbatore; HAU, Hisar; IISC, Bangalore and Defence Institute of Bio-energy Research (DIBER), Haldwani have taken initiatives to standardize the tissue culture protocol of Jatropha for commercial cultivation. Studies on morphogenesis, plant regeneration and shoot bud proliferation from auxiliary nodes (Sujatha and Mukta, 1996) and leaf sections of non-toxic *Jatropha curcas* (Sujatha et al. 2005) have been carried out at Directorate of Oilseeds Research, Hyderabad and plant regeneration from leaf explants derived embryogenic callus (Jyoti *et al*,2005) ; production of clonal plants from nodal segments (Rajore et al., 2002) and efficient plant regeneration via shoot tip explants(Rajore et al., 2002) at Rajasthan university. DIBER has identified two high yielding Jatropha germplasm viz., DARL-1 with 34.4% and DARL-2 with 36.5% oil content from Himalayan valley. The laboratory is exploiting this trait for mass multiplication through tissue culture to over come the problems in vegetative propagation and to take advantages of *in vitro* propagation owing to the problem i.e. Genetic uniformity being cross pollinated crop, nonavailability of seed certification standards, diseases and infection and perennial plant. Micro-propagation and genetic engineering of *Pongamia* have not received much attention. Mint Bio-fuels Ltd, Pune is engaged in developing tissue culture protocol of Pongamia for commercial production.

24.1 Technological Constraints

As against direct propagation through seed and cuttings there are certain technological constraints i.e. problems and advantages in vegetative and *in vitro* propagation. Certain advantages *in vitro* propagation over vegetative propagation is given below:

Vegetative Propagation – Problems	***In Vitro* Propagation - Advantages**
1. Time consuming for growth of mother trees. 2. Requirement of large land area 3. Quality of planting material affected by pests. 4. Transport difficulties- bulkiness of planning material.	1. True to type production. 2. Genetic uniformity. 3. Mass multiplication in small area. (in half an acre land it is possible to produce and supply about 5 million plants per year) 4. Low cost of production. 5. No seasonal and climatic dependence-regular supply. 6. Disease free plantlets.

24.2 Biosynthesis of Storage Oils

Plants store reserve material like oil in their seeds for the growth of the next generation. TAG is present in most plant organs including leaves, petals, fruits, anthers and developing seeds (Hobbs et al., 1999) and it is thought to be synthesized within the membrane of Endoplasmic Reticulum (ER) and subsequently released into the cytosol in the form of oil bodies, which bleb from cytoplasmic surfaces of the ER (Huang, 1992). In the mature seed, TAG is stored in densely packed lipid bodies that are roughly spherical in shape, with an average diameter of 1 pm (Huang, 1992; Murphy, 1993 and Herman, 1994). This size does not change during seed development, and accumulation of oil is accompanied by an increase in the number of lipid bodies. The membranes of isolated lipid bodies contain both phospholipids and major characteristic proteins, termed oleosins. However, these constituents are less than 5% of the weight of the oil body, with TAG constituting by far the major component (90 to 95%).

24.3 Kennedy Pathway

Seed tri acyl glycerol (TAG) biosynthesis is located in the ER with glycerol-3-phosphate and fatty acyl-CoA as the primary substrates. There are three acyltransferases and a phosphohydrolase involved in the plant storage lipid bioassembly, namely glycerol-3-phosphate acyltransferase (GPAT, EC 2.3.1.15), lyso-phosphatidic acid acyltransferase (LPAAT, EC 2.3.1.51), phosphatidate phosphohydrolase (PAPase, EC 3.1.3.4), and diacylglycerol acyltransferase (DGAT, EC 2.3.1.20). The three acyltransferases catalyze the stepwise acylation of the glycerol backbone out of which the final step being the acylation of *sn*-1, 2-diacylglycerols (DAGs) by DGAT to form TAGs, a biochemical process generally known as the Kennedy Pathway (Kennedy, 1961; Barron and Stumpf, 1962; Stymne and Stobart, 1987). The acyl-CoA dependent acylation of *sn*-1,2-DAG as catalyzed by DGAT is the only enzymatic reaction in the traditional Kennedy pathway that is exclusively committed to TAG biosynthesis. GPAT is the first committed enzyme in the glycerolipid biosynthesis (Pillai et al., 1998). In Brassicaceae, GPAT gene was only characterized from *Brassica juncea*. Kenedy Pantway is given below:

KENNEDY PATHWAY

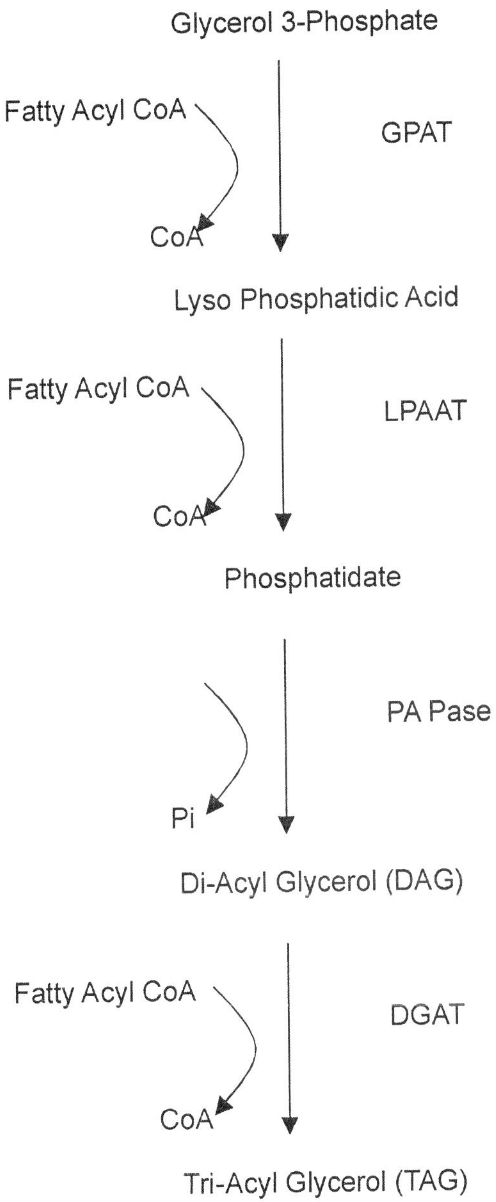

GPAT Glycerol-3-Phosphate Acyl Transferase
LPAAT Lyso-Phosphatidic Acid Acyl Transferase
PAPase Phosphatidate Phosphohydrolase
DGAT Diacyl-Glycerol Acyl Transferase

From the studies on assembly of three fatty acids onto a glycerol backbone it is known that straightforward Kennedy pathway does not always occur. However, in many oilseeds, most fatty acids produced in the plastid are not immediately available for TAG biosynthesis. Instead, the acyl chains enter into phosphatidyl choline, where they become desaturated or otherwise modified. In the first mechanism, a fatty acid attached to CoA and a fatty acid on PC may essentially trade places. Such an acyl exchange probably occurs by the combined reverse and forward reactions of an acyl-CoA:PC acyltransferase (Stymne and Stobart,1987). The resulting acyl-CoA may then be used as an acyl donor in TAG synthesis. The exchange reaction allows the 18:l, newly produced and exported from the plastid to enter PC, while desaturated or otherwise modified fatty acids depart for TAG or other lipids. The second mechanism by which PC can participate in TAG synthesis is by donation of its entire DAG portion (Stymne and Stobart, 1987). In many plants, the synthesis of PC from DAG and CDP-choline appears to be rapidly reversible as catalyzed by the CDP-choline phosphotransferase (Slack et al., 1983). The reversibility of this reaction allows the DAG moiety of PC, including any modified fatty acids, to become available for TAG synthesis because DAG is a common precursor to both membrane PC and storage TAG.

The basic metabolic pathways that lead to the synthesis of the major plant glycerolipids are now mostly understood. Much of the research effort in plant lipid biosynthesis over the past many years has been directed towards obtaining clones for enzymes in the pathway.

24.4 Modification of Seed oil Content

Vegetable oil is ubiquitous in everyday life, which could be an alternative to petroleum for chemical feedstock. Hence, increased seed oil content of oil seed crops such as canola, sunflower and soybean has been a target trait of plant breeders for many years. Engineering plants with increased flux through fatty acid synthesis to enhance oil content have been shown to be a difficult task due to complexity associated with regulatory mechanism. Up regulation of any single gene in the pathway have not always improved the oil yield. Although, over expression of a regulatory enzyme, ACCase in the rapeseed plastid caused a slight increment (about 3-5 %) in the oil yield (Roesler et al., 1997), expression of condensing enzyme KASIII in tobacco, rather than an increased fatty acid content, resulted 5-10 % decrease in fatty acid content (Dehesh et al., 2001) suggesting the important role of the regulatory factors in controlling the entire pathway.

1Over expression of the yeast long chain *sn*-2 acyltransferase resulted in more than 50% increase in seed oil content of *Arabidopsis* and rapeseed (Zou et al., 1997). In another study, seed specific over expression of *DGAT*1cDNA in wild-type *Arabidopsis* enhanced oil deposition and average seed weight, which were correlated with *DGAT* transcript levels. DGAT activity was increased by 10 to 70% in the seeds of transgenics. This shows the important role of DGAT in regulating the quantity of seed TAG and the sink size in developing seeds (Jako et al., 2001). Transformation of tobacco with At *DGAT* showed a marked increase (up to 7 times) of TAG content (Bouvier-Nave et al., 2000). Expression of *M. ramanniana DGAT2A* in insect cells also increased the total amount of TAG 2-3-fold in those cells. Together, these studies show that targeting the enzymes of the Kennedy Pathway especially DGAT may be very useful in increased oil yield of plants (Table 28).

Table 28: Genetically modified crops with improved oil contents.

S. No.	Gene	Increase in Oil Content (%)	Crop	Reference
1.	ACCase	3-5	Rapeseed	Roesler *et al.*, 1997
2.	Yeast sn-2 acyl transferase	8-48	Arabidopsis	Zou *et al.*, 1997
3.	*At* DGAT	9-12	Arabidopsis	Jako *et al.*,2001
4.	*At*DGAT	7times	Tobacco	Bouvier-Nave *et al.*, 2000
5.	*At*GPAT	8-29	Arabidopsis	Jain *et al.*, 2000
6.	Umbelopsis DGAT 2A	1.5% seed weight	Soybean	Lardizabal *et. al.*, 2008
7.	Yeast gpd 1	40.0% of lipid content	Brassica napus	Helene *et al.*, 2007

DIBER is also taking long strides for enhancing drought tolerance and oil content of Jatropha through genetic transformation (Fig 56). In an effort to transform Jatropha for enhanced oil content, regeneration protocol for the Jatropha has been established. The Napin promoter from *Brassica napus* and DGAT gene from *Arabidopsis thaliana* has been cloned (Fig. 55).

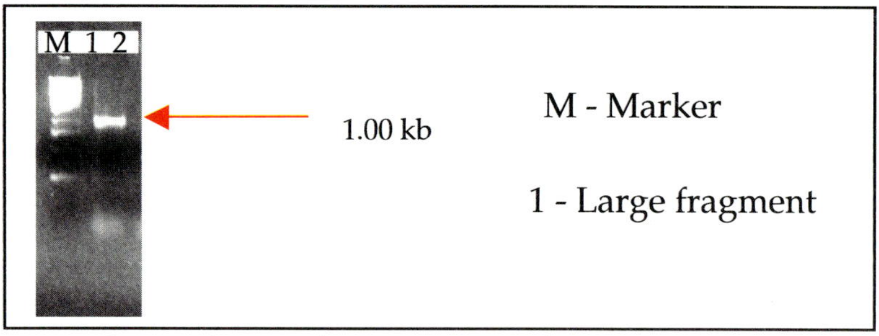

Fig. 55. Cloning of Napin promoter from *Brassica napus*

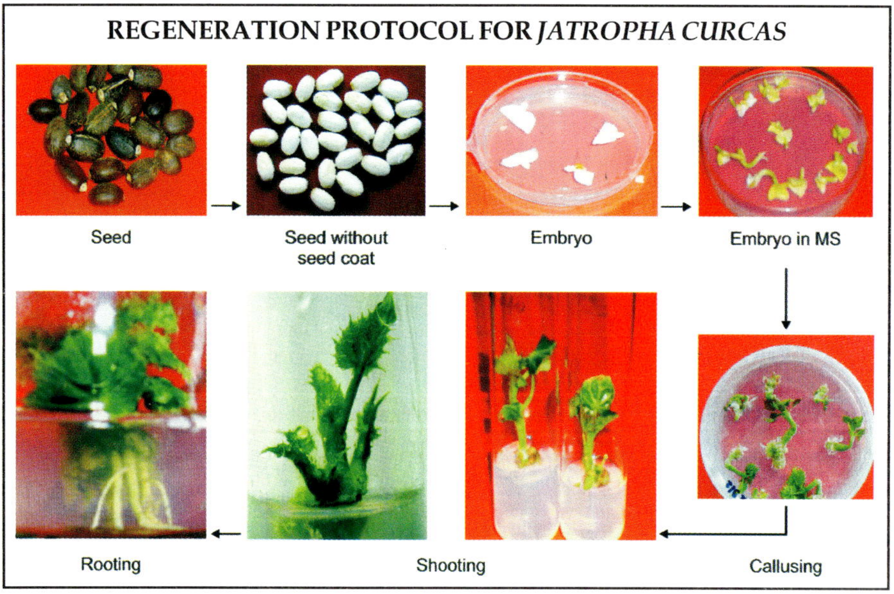

Fig. 56. *In Vitro* regeneration of Jatropha

25. DETOXIFICATION OF SEED CAKE

Seed cake is a rich source of protein and macro and micro nutrients. However, because of toxicity, it is not used as livestock feed. The toxic components in Jatropha cake are phorbol ester, trypsin inhibitor, alkaloids, glycocides, curcanoleic acid, curcin a phytotoxin similar to ricin in castor and HCN in young Sorghum leaves and *Tapioca rind* and purgative oil as in castor and croton oil. Although heat treatment or the combination of heat and chemical (NaOH and NaOCl) treatments are being tried to inactivate the above toxic components, yet no proven technology is presently available for detoxification. TNAU, Coimbatore and NBRI, Lucknow are engaged in heat and chemical treatment to neutralize the toxicity.

Reduction of curcin, a toxic substance content in the seed may be taken up by various techniques like (i) conventional mutation breeding supported by high throughout screening technique (ii). Knockout of curcin gene by homologous recombination's and (iii) Silencing curcin gene expression by RNA. However, curcin removal by these techniques are still in infancy.

26. COMMERCIAL PRODUCTION OF BIO-DIESEL IN INDIA

India produced 40 thousand tonnes of bio-diesel in the year 2005. About twenty one states are currently engaged in cultivation of Jatropha, Pongamia and other TBOs. Visible economic production is expected after 2-3 years. Operating cost of bio-diesel production is mainly dependent on cost of feed stock. Securing own feedstock to ensure supply at a fair price and sourcing it locally to avoid long haulage are critical factors in controlling profitability. The cost is Rs.15,000 - 20,000/ - per MT of bio-diesel produced, however, it may increase in future.

Few firms and NGO's engaged in commercial production or bio-diesel and fabrication of states, bio-diesel plants are:

- Chhatisgarh Bio-fuel Development Authority, Raipur
- Aatmiya Bio-fuels Pvt. Ltd., Vadodara, Gujarat
- Gujarat Oelo Chem Ltd., Panoli, Gujarat started biodiesel production from vegetable oil based feed stock since 2005.
- Bio-diesel Technologies, Kolkata
- Kochi Refineries Ltd, Kochi
- Shirke Bio-healthcare Pvt. Ltd., Pune
- Mint Bio-fuel, Pune has biodiesel plant of 400 L/day capacity from Pongamia.
- Aurangabad Bio-diesel Production Unit, Aurangabad (Maharastra)
- Bio-diesel technologies, Kolkata produces 450 L/day of bio-diesel.

- Diamond Energy Resources Pvt. Ltd work on Jatropha plantation.
- Sirk Bio-healthcare Pvt Ltd, Pune set up a refinery with the capacity to process 500 L bio-diesel/day from Jatropha oil. The refinery will also produce 1MW power with oil cake apart from natural gases.
- NOVA Bio-fuels Pvt Ltd plans to set up 30 TPD bio-diesel plant in Panipat.
- Natural Bio-energy Ltd is setting up 300 TPD bio-diesel plant in Kakinoda (A.P)
- Sagar Jatropha Oil Extractions Pvt Ltd, Vigaywada is setting up bio-diesel plant from Jatropha.

Industries - Reliance India Ltd, D1 Oil (a UK producer of green fuel), Mahendra & Mahendra , Godrej Agrovet, Emami Group and Natural Bio-energy Ltd. are venturing in a big way and are supposed to be the futuristic players in the field of bio-diesel in India. Few smaller Indian companies like Nandan Bio-agro and Lab India Bio-tech have also tied up with D1 oils to produce Jatropha bio-diesel.

27. INSTITUTION WORKING ON BIO-DIESEL IN INDIA

The institution and agencies which are engaged in this work include Punjab Agricultural University; Indian Institute of Technology, Delhi; Indian Institute of Technology, Chennai; Mahindra and Mahindra (M&M); Indian Institute of Science, Bangalore; Indian Institute of Petroleum, Dehradun; Indian Institute of Chemical Technology, Hyderabad; Indian Oil Corporation, Faridabad; Harbinsons Biotech, Gurgaon, IIT, Guwhati, DCE, Delhi , MINT Biofuel, Pune, Central Salt Marine and Chemical Research Institute (CSMCRI), Bhavnagar , GB Pant Institute of Agriculture & Technology, Defence Institute of Bio-Energy Research (DIBER), Haldwani and several other institutions.

28. PRODUCTION AND BLENDING OF BIO-DIESEL

As per European directives replacing 5.75% of diesel and petrol used in road transport with bio-fuel by 2010, the Govt of India has also announced its bio-fuel purchase policy in Jan 2006 to achieve 5% blending of conventional fuel. Accordingly, the Planning Commission has launched a National Mission on bio-diesel. It is estimated that to achieve 5% blending of bio-fuel, India need 2.6 million tonnes of oil.

India is witnessing and availing the advantages of the improvements, which have been taking place in the diesel engine technology including the automobile sector. These improved engines required bench-marked fuel otherwise their performance and life could be hampered. In India, we are also witnessing growing menace of adulteration in fuels including diesel. In these circumstances, it is desirable that final stage of processing and blending of bio diesel and its subsequent distribution/ marketing is carried under the supervision and control of oil companies so that the stringent conformance to specifications is observed and availability of quality fuels to the customers is ensured. The National Oil Companies have to play significant role in this regard. An appropriate model covering the operations, production of seeds, extraction of vegetable oil, production of crude bio diesel and final processing and blending of bio diesel need to be evolved with proper allocation of responsibilities (seed producers, oil / bio diesel processors and the oil-companies) and quality characteristics which need to be ensured at different stages of the operations to be set for standardization.

In India, Bureau of Indian Standards (BIS IS-1460 - 2000) is in the advanced stage of finalization of standard for B100 Bio diesel fuels. The same specifications shall be applicable for bio diesel for blending. Similarly, OEM approvals for blended and B100 bio diesel are in the process through pilot level trials under way at different places.

29. EVALUATION OF BIO-DIESEL

Experimental studies in IC engines have been conducted in Indian Institute of Technology, Kanpur (Aggarwal, 1998) and Chennai (Kumar et al., 2003). In majority of cases, performance evaluation of bio-diesel in IC engines has been done with 5% blend with very promising results. Only limited trials on B10 and B20 have been conducted for evaluation of engine health and emission. Railways have successfully run locomotive engines on 5% blends of bio-diesel in association with Indian Oil Corporation Ltd. HPCL is carrying out field trials with BEST, Mumbai. Daimler - Chrysler (DC), India completed field trials on two C-Class Mercedes-Benz C-220CDI cars with pure bio-diesel produced by CSMCRI, Bhavnagar and clocked over 5900 Km under hot and humid conditions without engine modification during April-May 2004. The laboratory has been playing its Tyota Qualis on neat bio-diesel since August 2006 and over 90000 km run hs been completed. This institute, in collaboration with DC India, carried out test run trial in Leh and also conducted emission tests jointly with ARAI, Pune with encouraging results. IIT Delhi is studying the performance of B20 in TATA Indica which has already covered more than 6000 km and shown significant benefits in emission. IIT, Guwahati recorded improvement in brake thermal efficiency and reduction in brake specific fuel consumption in engine specially at higher loads with B15 and B20 Pongamia and Jatropha bio-diesel in a 5 HP single cylinder engine. Delhi College of Engineering is testing TATA Indica with 20% blend for the last 1 year. Haryana, Maharashtra, Karnataka, Gujarat State Transport buses have been successfully run by using B5.

General Motors and the Daimler- Chrysler are monitoring the effects of B20 bio-diesel in real world fleet. Chrysler has put over 1,50,000 km in last 5 years while . General motors clocked 5 million km with 238 vehicles in two fleets. None of the fleets has experienced any engine problem with bio-diesel (Das and Madhav, 2006). Bureau of Indian standards (BIS) for quality control of bio-diesel has not yet been formulated. Bio-diesel producers widely use ASTM D 6751 and EN 14214 for which the testing facilities are available in the country. In India, implementation of Euro-III and Euro-IV fuel norms requires huge investment for upgrading the refinery. Bio-diesel also could be included in the programme to achieve the above norms. CSMCRI, Bhavnagar has produced Euro III bio-diesel from Jatropha. Evaluation of bio-diesel revealed that smog forming potential, unburnt hydrocarbons, carbon monoxide and sulphur dioxide emissions significantly reduced while NOx emission marginally increased but within acceptable limits. Detail analysis is given in Table 29 & 30.

Table 29. Average B100 And B20 Emissions (In Percentage)

Emission	B100	B20
Carbon monoxide	-48	-12
Total unburned hydrocarbons	-67	-20
Particulate matter	-47	-12
Sulphates	-100	-20
Air toxics	-60 to -90	-12 to -20

Table 30. CO and NOx emission in diesel and blended diesel

S. No.	Hours	Diesel Oil	Diesel+ J. Oil (20:80)
1	Oil consumption lit/hr	1.36	1.32
2	CO emission %	0.05	0.03
3	NO emission ppm	192	1400

Ignition delay was higher in neat *Jatropha* oil which increased further in blend and DFO while it was reduced with ester. Peak pressure and rate of pressure risk was higher with all methods compared to *Jatropha* oil operation. *Jatropha* oil and methyl ester showed higher diffusion combustion compared to standard diesel operation. However, dual fuel operation resulted in higher premixed combustion. They concluded that trans esterification of *Jatropha* oil and methanol induction can significantly enhance the performance of vegetable oil fuel engine. Diesel engine performance test using *Jatropha curcas* oil, ethyl ester was studied by Ouedraogo, et al. (1991) and observed that ethanol application reduced the viscosity. Fuel properties of raw oil and esterified product have been compared with other seed oils and diesel fuels. *Jatropha* oil ethyl ester produced 81% of the max power, 86% of the max to range and 115 % of the specific fuel consumption (SFC) rate of diesel.

Problems in using neat vegetable oils for engine performance, comparative performance of Jatropha oil and diesel and comparison of diesel, methyl and ethyl ester are given in Table 31, 32 & 33.

Table 31. Engine performance

Problems in Using Neat Vegetables Oils	Durability Problems
• Operational Problems	• Deposit formation
• Starting ability	• Carbonization of injector tip
• Ignition	• Piston ring sticking
• Combustion parameters	• Lube oil dilution
• Performance parameters	• Fuel filter plugging
• High viscocity	• Trans-esterification
• Extremely low volatility	
"Vegetable oil need to be transesterified for their conversion to bio-diesel"	

Table 32. Comparative properties of Jatropha oil and Diesel

Properties	J. Oil	Diesel
Viscosity (C. S.) 30°C	5.51	3.6
Density (gm/cm³) 4°C	0.92	0.85
Freezing point (°C)	2	-14
Cetane point	38	47-59
Combustion point (°C)	40	80
Carbon %	0.64	0.15
Sulphur%	0.13-0.16	1-1.2
Saponification value	188-198	-

Table 33. Comparison of Diesel, Methyl and Ethyl Ester of *Jatropha* Oil

Properties	Diesel	Methyl Ester	Ethyl Ester
Density (gm/cm³)	0.85	0.88	0.88
Combustion point (°C)	55	192	190
Viscosity (C. S.)	2.8	4.84	5.54
Calorific value (MJ/Kg)	45	41	42
Cetane No	47.5	52	59
Ester (%)	Nil	99.6	99.3
Sulphur	< 0.5	0.01	-
Carbon (%)	< 0.35	0.24	0.18

30. INITIATIVES MADE IN INDIA

India being mega bio-diversity country has produced bio-diesel from Jatropha and various successful trials using bio-diesel are listed below. However, there is big technology gap to make this venture commercially and eco-friendly viable.

- Indian Railway (IR) being the largest consumer of Petro diesel realized its importance a few years back and in association with Indian Oil Corporation conducted a trial of running New Delhi-Amritsar Shatabdi Express on 31 Dec 2002 to assess its suitability with a 5% blend with Petro diesel which went on smoothly.
- The Jan Shatabdi Express between Lucknow and Allahabad was run with a 10% blend of Bio-diesel for 3 round trips (404 Km each) during June 2004 and all these runs went on smoothly.
- In Indian Railways Loco Engine Test/Trials was conducted on 5%, 10% and 20% Bio-diesel blends on 16 Cylinder Alco Engine

(3100 HP) at Research Designs and Standards Organization (RDSO), Lucknow and Horse Power, Firing pressures and Exhaust Gas temperature were observed satisfactory.

- BEST, MUMBAI has conducted successful trials with B5, B10 & B20 blends on their bus fleet.
- Maharastra, Karnataka and Gujrat state transport buses have successfully done using B_5.
- IOC, Haryana Roadways has been running 20 Buses from April 2004 on 5% Bio-diesel-Diesel Blends & another 20 Buses are being run on petro-diesel as reference. No significant difference in fuel consumption was observed on buses with or without bio-diesel.

31. COST BENEFIT ANALYSIS/ SPIN-OFF BENEFITS

Since Jatropha plantation incurs one time investment with routine care and plant sustain for 40-50 years with fruit production thus there is no requirement of repeated expenditure as compared to traditional cropping system. Jatropha plant produces fruits for long period and thus bio-diesel will be obtained accordingly.

Development of genetically modified Jatropha with oil content enhancement and reduced curcin level may help to future renewal of Jatropha plantation. Bio-diesel is eco-friendly, non-toxic, economic, produced from renewable agriculture based source of energy and compatible with current CI engines and it can reduce import of petro diesel, saving foreign exchange and reduction of carbon load from the environment.

32. FOCUS

1. Establishment of Technologies
    - Selection and development of superior germplasm/cultivars for different agro-climatic zones based on evaluation of genetic stock in different agro-climatic zones and application of genetic engineering to increase the seed yield potential and oil content.
    - Rapid Micro propagation techniques to develop true to type plants from elite germplasm.
    - Development of location specific agro-technologies and inter cropping pattern for better economic returns.

- Development of commercially viable trans-esterification process with minimum energy and water consumption and zero effluents discharge having future scope of upscaling bio-diesel yield.
- Enhancement of shelf life of bio-diesel with stabilizers, anti-freezing and anti-oxidant agents.
- Detoxification of oil cake and its conversion into protein rich feed for livestock.
- Methanation of oil cake for bio-gas power and industrial use.
- Conversion of methanation sludge as bio-fertilizer.

2. Establishment of Bio-fuel Park for pilot production of bio-diesel as per standards.
3. Performance and pollution evaluation of bio-diesel in defence vehicles and to minimize the emission of NOx.

33. COMPETENCE LEVEL OF DRDO LABS

DIBER

The proficiency of DIBER has been in the development of state of art technologies in the field of Agro-horticulture and release of a number of vegetable varieties/hybrids. The lab also has competence in plant genetic engineering since development of transgenic in various vegetable crops for cold and draught tolerance is in progress.

VRDE

VRDE is working on research, design, development and technical trials and evaluation of all types of wheeled and light tracked vehicles of combat and specialist roles. VRDE is certification lab for engine testing for forces and well equipped with all facilities required for fuel testing.

DMSRDE

DMSRDE has got the privilege to be the only lab in DRDO to look after R&D work pertaining to development of Fuels and Lubricants for all wings of Defence forces. A large number of developed and indigenized products have been introduced in the services. DMSRDE has got the bench testing and evaluation facilities required for all types of Fuels and Lubricants.

DRDE

DRDE has got expertise in the field of toxicological studies and has been working on Ricin, an analogue of curcin for many years. The lab has competence to detoxify the phorbol esters and curcin in cake & make it suitable as livestock feed. After the extraction of oil, remaining 70% percent is the cake which is toxic due to the presence of curcin (a toxic protein present in the seeds), trypsin inhibitor, HCN, phorbol esters, alkaloids, glycosides and curcanoleic acid. The cake can not be fed to animals as such due to toxic and anti-nutritive substances.

DIHAR

DIHAR, a unique lab of DRDO situated in cold desert, has competence in High Altitude Research Technologies. The lab has strength to undertake bio-fuel storage studies in cold desert in association with DMSRDE and vehicle testing in association with EME and VRDE.

DRL

DRL, the only DRDO lab situated in North East, has competence in plant bio-technology, specially tissue culture which can be exploited in screening of Jatropha germplasm collected from NE region and mass multiplication of suitable cultivar for seed production. In addition, the lab has embarked on micro-algae for bio-diesel production and has screened some algal strains of NE region. Purification of culture for multiplication is under progress.

contd....

34. JATROPHA - ACTION - USES

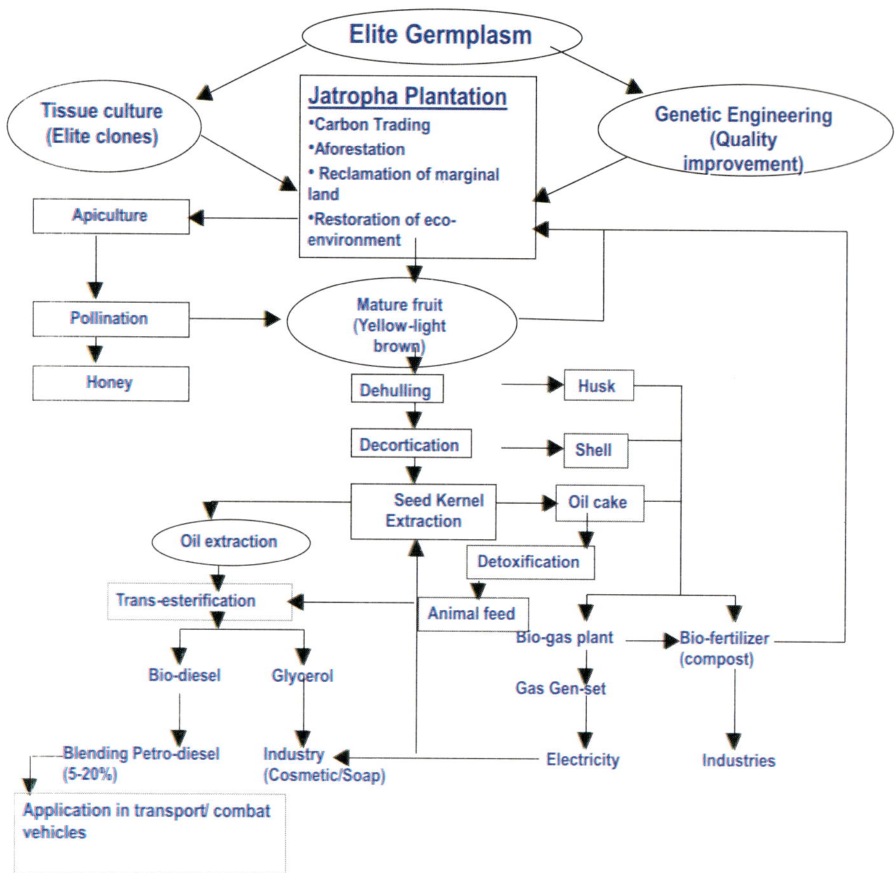

contd....

35. PROMINENT WORKERS ON BIO-DIESEL IN INDIA

There are a large number of Scientists/researchers/academicians/ NGO's who are actively engaged in bio-diesel production through *Jatropha / Pongamia* cultivation, its evaluation, marketing, training etc. Few of them are listed below:

1. Dr. D. N. Tiwari, Vice Chairman, State Planning Board, Govt. of Chattishgarh.
2. Dr. R. S. Kureel, Director, NOVOD Board, MAO, GOI, 86, Sector-18, Institutional area, Gurgaon-122015.
3. Dr. M. Paramathma, Prof. and Head, Deptt of Tree Breeding Forest College and Research Institute, TNAU, Mettupalayam, Coimbatore-641301.
4. Shri Sakeel Ahmed, DRM, South-Eastern Railway, Kharagpur, West Bengal.
5. Dr. Renu Swarup, Director, Deptt of Bio-Technology CGO Complex, Block No. 2, 7th Floor, Lodhi Road, New Delhi-110003.
6. Dr. S. B. Lal, Dean, College of Forestry, Allahabad Agriculture Institute, (Deemed University) Allahabad.
7. Dr. Naresh Kaushik, Manager, Indian Oil Bhawan,1.Sri Aurobindo marg, Yusuf Sarai, New Delhi-110016.
8. Dr. V. K. Gour, Associate Prof., Deptt of Plant Breeding and Genetics, JNKVV, Jabalpur.
9. Dr. Arvind Shukla, Associate Prof., deptt of Plant Breeding College of Agriculture, G.B. Pant University, Pantnagar, Udham Singh Nagar, Uttarakhand-263145.
10. Dr. Laljeet Singh, Associate Prof., Deptt of Forestry, Indira Gandhi Agriculture University, Raipur, Chattishgarh-492006.
11. Dr. D.K. Tuli, CEO, Indian Oil Technologies Ltd. IOC Ltd. (R&D Centre) Sector-13, Faridabad, Haryana-121007.
12. Shri G. M. Pillai, World Institute of Sustainable Energy (WISE), Surya-Suman, Room No. 2, Kalyani Nagar, Pune-411006.
13. Shri O.D. Sharma, RHS Biofuels Pvt. Ltd. A-82, Sector 14, Noida.
14. Shri R.K. Chaturvedi, Project Officer, Chattisgarh, Biofuel Development Authority 'Yasosiddh Bhawan', Near Masonet-48, Sector-1, Shankar Nagar, Raipur.

15. Dr. S. D. Singh, Vice Chairman, Biofuel Development Board, Dehradun, Uttarakhand.
16. Smt. Veena Sekhri, Chairman, Uttaranchal Bio-fuel Development Board D-110, Sector-4, Defence Colony, Dehradun.
17. Shri. P. Venkateshwar Reddy, Director, Roshini Bio-Tech Pvt. Ltd.
18. Shri. A.K. Goel, Director, PCRA, Ministry of Petroleum & Natural Gas, GOI, New Delhi.
19. Shri. Shailendra Jain, Director, Aditya Biotech Lab & Research Pvt. Ltd., Near Nalghar Chowk, Chhotapara, Raipur.
20. Dr. Ganapathy Arumugam, Managing Director, Enhanced Bio-fuel & Technologies (I) Pvt. Ltd., Biofuel Research & Development Centre, 5/10-C, Alankar Garden, G. N. Mills Po, Coimbatore-641029, India.
21. Shri. Raju Mansinghka, Director, Ministry of Agriculture, CEO in Mansi Oils and Grains Pvt. Ltd.
22. Dr. D. K. Khare, Director, MNES, New Delhi.
23. Dr. V. Ranga Rao, Formerly Director, (Oil Seeds), Dte. of Oil Seeds Research, (ICAR), Rajendranagar, Hyderabad-500030. Currently, ETV Annadata, Ramoji Film City, Hyderabad-501512.

36. FURTHER READINGS

1. Anonymous (2005) Bio solid and liquid managemet using biosanitizer based stabilization process and methanation (rsanthanam delhi @ yahoo.com).
2. Anonymous (2006) Excerpts from report "Auto fuel doping to be mandatory from,12". The Economic Times, 5th Jan 2007.
3. Anonymous (2006) Excerpts from the report "Bio-fuels development board on the cards". The Economic Times, 26th Dec, 2006.
4. Ajiwe,V.I.E., Okeke. C.A. Agbo, H.U., Ogunleye, G.A. and Ekwuozor, C.C. (1996). Extraction, characterization and industrial uses of velvet tamarind, physic nut and nicker nut seed oil. *Bioresource Technology*, 57:297-299.
5. Abigor, R.D. and Uadia, P.A (2001). Lipid composition of *Jatropha curcas* L seed oil. *Rivista Italiana Delle Sostanze Grasse*, 78 : 163-165.

6. Agrawal, A.K.(1998) Vegetable oils versus diesel fuel: development and use of biodiesel in a compression ignition engine. *TIDE*, 83: 191-204.

7. Akinttayo, E.T. and Bayer, E. (2002). Identification of oils by NMR spectroscopy. *Rivista Italiana Delle Sostanze Grasse*, 79 : 207-210.

8. Aregheore, E.M.,Becker, K. and Kakkar, H.PS. (2003). Detoxification of a toxic variety of *Jatropha curcas* using heat and chemical treatment and preliminary nutritional evaluation with rats. *S Pac.J.Nat.Sci.* 21: 50-56.

9. Arif, M. and Sharma, S. and Das S.C. (2007). Incidence of leaf minor on *Jatropha curcas*- a bio-diesel plant in Secunderabad. *J. Exp.* 2007. *India* 10 (1) : 107.

10. Banerjee, R; Chowdhury, A.R; Misra ,G; Sudarsanam, G; Verma, S.C and Srivastava, GS (1985). Jatropha seed oils for energy. *Biomass*, 8: 277-282.

11. Baron, E.J. and Stumpf, P.A. (1962). Fat metabolism in higher plants. XIX. The biosynthesis of triglycerides by avocado-mesocarp enzymes. *Biochim. Biophys. Acta.* 60 : 329-337.

12. Bhattacharya, P.K., Shukla, A, Singh, J. and Pandey, R.D. (2005). *Jatropha* training Schedule. Rastriya Tilhan Avam Vanaspati Tale Vikas Board Gurgaon and GBPUA&T, Pantnagar (UA). 1-31.

13. Bhattacharya, P., and Joshi,. B (2006). Strategies and institutional mechanism for large scale cultivation of *Jatropha curcas* under agro-forestry in the context of proposed bio-fuel policy of India, Envis. *Bull. Grassland Ecosystem and Agro-forestry*, 1 (2): 58-72.

14. Bhojvaid,P.P (2006) .Raising large-scale Jatropha plantation in India: present status and future prospects. Biofuels:towards a greener and secure energy future, 281p.

15. Bouvier-Nave, P., Benveniste, P., Oelkers, P., Sturley, S.L. and Schaller, H. (2000). Expression in yeast and tobacco of plant cDNAs encoding acyl CoA:diacylglycerol acyltransferase. *Eur. J. Biochem.* 267 (1) : 85-96.

16. Calvin, M (1979). In- A text book of Plant Biotechnology, by R.C. Dubey (1995) S. Chand and Co Lt. Ramnagar, Delhi 110055, 311 - 313, *Bioscience* 29, 573-78.

17. Carraretto, C., Macor, A., Mirandola, A., Stoppato, A. and Tonon, S. (2004). Biodiesel as an alternative fuel : Experimental analysis and energetic evaluation. *Energy*, 29: 2195-2211.

18. Chatakananda, P., Sriroth, K. Vaysse, L. and Liangprayoon, S. (2005). Fatty acid composition and properties of Jatropha seed oil and its methyl ester. Proceedings of 43rd Kasetsart Univ Annual Conference, Thailand, Feb 1-4: 641-648.

19. Chengappa,R. (2008). Global warming: What India should do.TOI,July:37-34

20. Chaturvedi, R.K. (2006). Biofuel development programme in Chattisgarh. Paper presented in conference. Conference organized at Rastrapathi Nlayam , Bolaram, Hyderabad on 9-10 June 2006.

21. Chavanne,C.G.(1938).Belgian Patent 422,877, August 31, 1937: *Chem. Abs.* 32:4313

22. Chivandi, E; Mtimuni, J.P; Read, J.S. and Makuza, S.M (2004). Effect of processing method on phorbol ester concentration, total phenolics, trypsin inhibitor activity and the proximate composition of the Zimbabwean *Jatropha curcas* provenance, a potential livestock feed. *Pakistan Journal of Biological Sciences*, 7: 1001-1005.

23. Dadwal,S.R.(2003). Kyoto Protocol and the future of Carbon Trading. The Financial Express, 27, October.

24. Das, L.M and Madhav, Venu. (2006). Research issues in development of bio-diesel engines. In : Biofuels Towards a Greener and Secure Energy Future by Bhojvaid, P.P Ed., p198.

25. Datta, S.K and Pandey,R.K. (1993). *Jatropha curcas* - a promising crop for new source of fuel. *Applied Botany Abstracts*, 13: 108-118.

26. Dehesh, K., Tai, H., Edwards, P., Byrne, J. and Jaworski, J.G. (2001). Overexpression of 3-ketoacyl acyl carrier protein synthase III in plants reduces the rate of lipid synthesis. *Plant Physiol.* 125: 1102-1114.

27. Diesel, R. (1912). *The diesel oil engine, Engineering* 93:395-406.

28. Diesel, R., (1913).The Diesel Oil-Engine and Its Industrial Importance Particularly for Great Britain, *Proc. Inst. Mech.Eng.*: 179–280 (1912). *Chem. Abstr.*7:1605.

29. Foidl, N., Foildl, G., Sanchez, M., Mittelbach, M. and Hackel, S. (1996). *Jatropha curcas* L as a source for the production of biofuel. *Bioresource Technology*, 58: 77-82.

30. Gaur, V.K. (2006) Production Practices including post harvest management of *Jatropha curcas.* Proceedings of bio-diesel conference towards energy independence- focus on Jatropha Ed. Brahma Singh, R Swaminatham and V Ponraj, Rastrapati Bhawan, New Delhi (2006).

31. Ginwal, H.S., Phartyal, S.S., Rawat, P.S., Srivastava, R.L. (2005). Seed source variation in morphology germination and seedling growth of *Jatropha curcas* Linn in central India - *Silva - Genetica,* 54 (2), 76 - 80.

32. Ghotra, B.S., Dayal, S.D. and Narine,S.S.(2002) Lipid shortenings: A Review. *Food Res.Інternl.J.* 35:1015-1048

33. Grimm, C. (1996). Manejo Integrade Plagas, 42: 23-30.

34. Grimm, P.C., and Maes, J.M. (1997), *Rev. Nicaraguense Entomol,* 39:13-26.

35. Grimm, C. and Guharay, F. (1998) *Bicontrol Sci. Technol,* 8: 365-376.

36. Gubitz, G M., Mittelbach, M. and Trabi,M. (1999). Exploitation of the tropical oil seed plant *Jatropha curcas* L. *Bio-resource Technology* (UK), 67:73-82.

37. Gunstone, F.D.(1986) Fatty acid structure. In: The lipid hand book. Ed.F.D. Gunstone., J.L. Hardwood., F.B., Padley. New York: Chapman and Hall. pp.1-24.

38. Gunstone, F.D.(2001) Soybean pace boost in oil seed production. *Inform.*11:1287-1289.

39. Hardwood, J.L.(1980) Plant acyl lipids: structure, distribution and analysis. In: The Biochemistry of plants,vol.4,P.K. Stumpf (ed) New York: Academic Press. Pp.2-56

40. Hilditch, T.P. and Williams, P.N. (1964). The chemical constituents of natural fats. London Chapman and Hall. p 745.

41. Herman, E.M. (1994). Cell and molecular biology of seed oil bodies. *In* : Seed Development and germination. H. Kigel. and G. Galili, (eds) New York : Marcel Dekker. pp. 195-214.

42. Hobbs, D.H., Lu, C. and Hills, M.J. (1999). Cloning of a cDNA encoding diacylglycerol acyltransferase from *Arabidopsis thaliana* and its functional expression. *FEBS Letter.* 452 : 145-149.

43. http://www.nabard.org/roles/ms/fw/pongamia.htm: Karanj for waste land Pongamian pinnata.

44. htpp:// www.vkrao.com/biotech.htm

45. htpp: // www.astm.org./ABOUT ASTM.html

46. httpp://www.jatropha/faq.htm.

47. Huang, A.H.C. (1992). Oil bodies and oleosins in seeds. *Annu. Rev.Plant physiol. Plant Mol. Biol.* 43 : 177-200.

48. Jako, C., Arvind, K Wei, Y., Zou, J., Barton, D.L. and Giblin, E.M. (2001). Seed specific overexpression of an *Arabidopsis* cDNA encoding a diacylglycerol acyltransferase enhances seed oil content and seed weight, *Plant Physiol.* 126 : 861-874.

49. Jesus, A. M.S., Pasqual, M., Dutra, L.F. and Chagas, E.A. (2003). *In vitro* culture of Jatropha zygotic embryos. Revista Ceres (Brazil), 288:183-189.

50. Joshi, Vandana. (2005). Cultivation of nutritional oil seed plant - *Jatropha curcas* for utilization of forest waste lands. *Annals of Forestry*, 13: 54-52.

51. Jyoti, Sardana; Amla,Batra; Ali, D.J (2000). An expeditious method for regeneration of somatic embryos in *Jatropha curcas* L. *Phytomorphology*, 50: 239-242.

52. Kantwal, R.P.S. and Soni, P.L. (2003). Biofuels. An opportunity for socio economic development and cleaner environment. India *Forester*, 129 (8), 939 - 949.

53. Kausik, N. (2006). Quality planting material and seed standards in *Jatropha curcas.* Proceedings of bio-diesel conference towards energy independence- focus on Jatropha Ed. Brahma Singh, R Swaminatham and V Ponraj, Rastrapati Bhawan, New Delhi (2006).

54. Kennedy, E. (1961). Biosynthesis of complex lipids. *Fed Proc. Amer.Soc.Exp.Biol* 20 : 934-940.

55. Knothe, G. (2001). Historical perspectives on vegetable oil based disease fuels. *Inform.* 12:1103-1107.

56. Kureel, R. S. (2006) . Prospects and potential of *Jatropha curcas* for bio-diesel production. In: Proceedings of bio-diesel conference towards energy independence- focus on Jatropha,p 43-74.

57. Lin,Juan ; Zhou , Xuan. wei ; Tang, Ke.Xuan and Chen,Fang (2004). A survey of the studies on the resources of *Jatropha curcas*. *Journal of Tropical and Subtropical Botany,* 12 : 285-290.

58. Makkar, H.P.S., Aderibigbe, A.O., Beeker, K. (1996). Comparative evaluation of non toxic and toxic varieties of *Jatropha curcas* for chemical composition digestibility, protein degradability and toxic factors. *Food chemistry* (United Kingdom), 62 (2), 207 - 215.

59. Makkar - H.P.S., Becher, K., Sporer, Fr. and Wink, M. (1997). Studies on nutrifire potential and toxic constituents of different provenances of *Jatropha curcas*. *Journal of Agriculture and Food Chemistry,* 45 (8) 3152 - 3157.

60. Metting, F.B.(1996). Biodiversity and application of mico algae. *J.Ind.Microbiol Biotechnol.* 17:477-489.

61. Martinez, S. J., Pedraza, R.M. and Garcia, Y. (2004). Influence of foliage drying method and the extraction solvent on the quantification of total extractable poly pherols. *Postos – y- Forrajes* 24 (4): 353-356.

62. Modi, N.K., Reddy, J.R.C., Rao, B.V.S.K. and Prasad, R.B.N. (2006). Lipase mediated transformation of vegetable oil into biodiesel using propan 2 ol as acyl accepter. *Biotechnology Letters,* 28: 637-640.

63. Mujumdar, A.M. and Misra, A.K. (2004). Anti inflammatory activity of *Jatropha curcas* roots in mice and rats. *Journal of Ethno-pharmacology,* 2004, 90 (I), 11- 15.

64. Meshram, P.B and Joshi, K.C. (1994). A new report of *Spodoptera litura* Fab Boursin (Lepidoptera: Noctuidae) as a pest of *Jatropha curcas* L. *Indian Forester* 120 (3): 273-274.

65. Muller,M.S. and Mechler, E. (2005). Medicinal plants in tropical countries. Traditional use experience facts. In: Medicinal Plants In Tropical Countries,168p.

66. Murphy, D.J. (1993). Structure, function, and biogenesis of storage lipid bodies and oleosins in plants. *Prog. Lipid Res.* 32 : 247-280.

67. Murphy, D.J. (1996). Engineering oil production in rapeseed and oil crops. *Trend Biotech.*14:206-213

68. Murphy, D. J. and Vance, J. (1999). Mechanisms of lipid body formation. *Trends Biol. Sci.* 24:109-115

69. Nemathy, E. K., Otre, J. W. and Calvin, M. (1980). Hydrocarbons from *Euphorbia lathyris.* Reprint 11351. Lawrence Berkeley Lab. University of California, California

70. Nitske, R, W. R and C.M., Wilson (1965). Rudolf Diesel. Pioneer of the age of power, University of OKLA HOMA Press, Novman: OKLAHOMA.

71. Ohlrogge, J.B. (1994). Design of new plant products: Engineering of fatty acid metabolism. *Plant Physiol.* 104 : 821-826.

72. Ohlrogge, J.B. and Browse, J.(1995). Lipid biosynthesis. *Plant Cell.* 7:957-970.

73. Parmathama, S., Parthiban, K.T., Balasubramanian, A.Venkatachalam, P. and Raghuram, P.(2005). Cultivation and management of Jatropha plantation. Forestry Series, TNAU Coimbatore, 218 p.

74. Parmathama, S., Reeja, K.T., Parthiban, D. and Malarvizhi (2006) Development of interspecific hybrids in Jatropha, Paper presented in conference. Conference organized at Rastrapathi Nlayam, Bolaram, Hyderabad on 9-10 June 2006.

75. Pillai, M.G., Certik, M., Nakahara, T. and Kamisaka, Y. (1998). Characterization of triacylglycerol biosynthesis in subcellular fractions of an oleaginous fungus, *Mortierella ramanniana* Var. angulispora. *Biochim. Biophys. Acta* 139 : 128-136.

76. reentech.htme\Biofuel Military\U_S_ DoD funds bio-fuel development for military Inside

77. Pak, M. and Allexi, A. (1994). Testing engine fuelled with physic oil. Landtechnik (Germany), 49:29-30.

78. Rao, R.V. (2006). Jatropha hype promise and performance. Proceedings of bio-diesel conference towards energy independence-focus on Jatropha Ed. Brahma Singh, R Swaminatham and V Ponraj, Rastrapati Bhawan, New Delhi (2006).

79. Rajore,S; Sardana, J. and Batra , Amla (2002). In vitro cloning of *Jatropha curcas* L. *Journal of Plant Biology*, 29: 195-198.

80. Rajore, Shilpa and Batra, Amla (2005). Efficient plant regeneration via shoot tip explant in *Jatropha curcas* L. *Journal of Plant Biochemistry & Biotechnology*, 14: 73-75.

81. Ratree, S. (2004). A preliminary study on physic nut (*Jatropha curcas* L) in Thailand. *Pakistan Journal of Biological Sciences*, 7:1620-1623.

82. Ravindranath - Nasi, Venka taiah - Bollu, Ramesh, Chimmani, Jaya Prakash - Pagadala and Oass, Biswanath (2003). *Jatrophane*, a novel macrocyclic bioactive diterpene from josyfolia. *Chemical and pharmaceutical. Bulletin* - Tokyo 51 (7), 870-871.

83. Roesler, K., Shintani, D., Savage, L., Boddupalli, S. and Ohlrogge, J.B. (1997). Targetting of the Arabidopsis homomeric acetyl-coenzyme A carboxylase to plastids of rapeseeds. *Plant physiol.* 113 :75-81.

84. Rong, Fang ; Wang ,Sheng .Hua ; Zhao,Xiao. Guang ; Guo, Liang; We. Qin; Tang, Lin ; Xu, Ying and Chen,Fang (2005). Identification of curcin by western-blot in calli generated from explants of *Jatropha curcas* L. *Sichuan Daxue Xuebao (Ziran-Kexueban)*, 42: 206-209.

85. Sangha M.K. , Gupta, P.K. ,Thaper, V.K. and Verma,S.R. (2004). Storage studies on plant oils and their methyl esters. Agricultural Engineering International,6: Manuscript EE 03 005.

86. Schwab, A.W., M.O. Bagby, and B. Freedman, (1987).Preparation and Properties of Diesel Fuels from Vegetable Oils, *Fuel 66*:1372-1378

87. Saturnino, H.M., Pacheco, D.D., Kalkida, J., Tominaga, N. and Goncalves, N.P.(2005). Cultivation of *Jatropha curcas* L. *Informe Agropecuario*, 26: 44-78.

88. Shah, S., Sharma, Aparna. and Gupta, M.H. (2005). Extraction of oil from *Jatropha curcas* L. seed Kernels by combination of ultrasonication and aqueous enzymatic oil extraction. *Bio resource technology*, 96 (1). 121 - 123.

89. Shanker, C. and Dhyani, S.K., (2006). Insect pests of *Jatropha curcas* L. and the potential for their management. *Current Science*. 91 (2) : 162-163.

90. Sharma, N. (2006). The Jatropha experience: Andra Pradesh, Paper presented in conference. Conference organized at Rastrapathi Nlayam , Bolaram, Hyderabad on 9-10 June 2006.

91. Sheeham, J., Cambreco, V., Duffield J., Garboski, M.M. and Shapouri, H. (1998). An overview of bio-diesel and petroleum diesel life cycles. A report by US Department of Agriculture and Energy,1-35 (*Journal of Scientific &. Industrial Research*,64:888).

92. Sheeham, J, Dunahay, T, Benemann, T et al.(1998). Alook back at the U.S. Department of Energy's Aquatic species programme: Biodiesel from algae. Clox out report NREL/TP-580.

93. Shukla, A. and Negi, K.D. Singh (2005). *Jatropha* (Ratan jot) Agricultural methods and uses. Directorate of Publication GBPUA&T, Pantnagar (U.A.).

94. Slack, C.R, Campbell, L.C, Browse, J.A. and Roughan, P.G. (1983). Some evidence for the reversibility of choline phosphotransferase catalyzed reaction in developing linseed cotyledons *in vivo. Biochem. J.* 263: 217-228.

95. Singh, B., Swaminathan, R and Ponraj, V. (2006). Bio-diesel conference towards energy independence -focus on Jatropha, Paper presented in conference. Conference organized at Rastrapathi Nlayam , Bolaram, Hyderabad on 9-10 June 2006.

96. Song, D., Fe, J. an Shi, D. (2008). Exploitation of oil-bearing microalgae for biodiesel. *Chinese J. of Biotechnology*, 24 (3): 341-348.

97. Spolaore, P., Jonnis- Cassam, C, Duran, E. et al. (2006) Commercial applications of microalgae. *J. Bioscience Bioengineer*, 101: 87-96.

98. Subramanian, K.A., Singhal, S.K., Saxena, M. and Singhal, S. (2005). Utilization of liquid bio-fuels in automotive diesel engine. An Indian perspective. *Biomass & Bioenergy*, 29: 65-72.

99. Sujatha, M. and Mukta, N. (1996). Morphogenesis and plant regeneration from tissue cultures of *Jatropha curcas. Plant Cell Tissue and Organ Culture* (Netherlands) 44 : 135-141.

100. Swarup, R. (2006). Quality planting material and seed standards in Jatropha, Paper presented in conference. Conference organized at Rastrapathi Nlayam , Bolaram, Hyderabad on 9-10 June 2006.

101. Sujatha, M; Makkar, H.P.S and Becker, K. (2005). Shoot bud proliferation from auxillary nodes and leaf sections of non-toxic *Jatropha curcas* L. *Plant Growth Regulation*, 47: 83-90.

102. Stienswat,W., Rattanaserrikiat, S. and Nimunchat, S. (1986). The studies on growing of physic nut (*Jatropha curcas*) in the large test plot. Research Report on Growing of Physic nut, Kesetsart Univ, Bankok,1:1-13.

103. Stienswat,W., Huangnikon, P. and Kamphukun, M. (1986). Trans-esreification of *Jatropha curcas* for diesel fuel substitute and lighting oil. Kasetsart Univ Fuel Project Research Report, 2:27-52.

104. Staubmann, R.; Foidl, G.; Foidl, N.; Gubitz,G..M.; Lafferty, R.M.; Arbizu,V..M..V. and Steiner, W. (1997). Biogas production from *Jatropha curcas* press cake. *Applied Biochemistry and Biotechnology* (USA), 63/65 p. 457-467.

105. Stymne, S., and Stobart, A.K. (1987). Triacylglycerol biosynthesis.*In* : *The Biochemistry of plants*,Vol.9,P.K.Stump and E.E.Conn, (eds) New York : Academic Press. pp 175-214.

106. Teixeira, L.C. (2005). Potential of oils for the production of biodiesel. *Informe Agropecuario*, 26: 18-27.

107. Thelen, J.J. and Ohlrogge, J.B.(2002). Metabolic engineering of fatty acid biosynthesis in plants. *Met Eng*.4:12-21.

108. Tuli, D.K. (2006). Bio-diesel production through trans esterification. Proceedings of bio-diesel conference towards energy independence focus on Jatropha Ed. Brahma Singh, R. Swaminatham and V Ponraj, Rastrapati Bhawan, New Delhi (2006).

109. Thitithanavanich, B. (1984). The effect of diameter and length of stem cuttings on root production of Physic nut. Bankok:17 leaves.

110. Tzen, T.C., Cao, Y.,L aurent, P. Ratnayake, C. and Huang, H.C. (1993). Lipids ,proteins and structures of seed oil bodies from diverse species. *Plant Physiol.* 101: 267-276

111. Van de Loo, F.L.,Fox, B.G. and Somerville, C. (1993). Unusual Fatty Acids. In.:Lipid Metabolism in Plants, T.S. Moore, Jr.,(ed) Boca Raton, FJL: CRC Press. pp 91-126.

112. Walton, J., (1939). The Fuel Possibilities of Vegetable Oils, *Gas Oil Power* 33:167–168 (1938). *Chem. Abstr.*.33:19396

113. Wei, Qin; Lu,Wei. Da ; Liao, Yi ; Pan, Shu. Lin ; Xu, ying ; Tang, Lin and Chen, Fang (2004). Plant regeneration from epicotyl explant of *Jatropha curcas. Zhiwu Shengli Yu Fenzi-Shengwuxue Xuebao*, 30: 475-478.

114. www. pcra.org/rnd/purchase.htmBiodiesel purchase policy (2005). Ministry of Petroleum and Natural Gas, Government of India.

115. www.biodiesel.org National Bio-diesel Board (US)

116. www.eia.doe.gov/oiaf/analysispaper/biodiesel/index.html Biodiesel performance, costs and use.

117. (www.biofuelsystems.com Biodiesel technology and alternative energy specialities. Biofuel System Ltd).

118. (www.powerservice.com). Diesel additives. Power Service Products

119. (www.planetark.com/dailynewsstory.cfm/newsid/30888/story.htm.) France opens second phase of bio-fuels plan.

120. (www.lubrizol.com) Lubrizol Corporation, an innovative speciality chemical company.

121. Wiebe, R., and J. Nowakowska , (1949). *The Technical Literature of Agricultural Motor Fuels, USDA Bibliographic Bulletin No. 10,* Washington, DC, pp. 183–195

122. Winkler,E., Foidl, N., Gubitz, G.M., Staubmann, R. and Steiner, W. (1997). Enzyme supported oil extraction from *Jatropha curcas* seed. *Applied Biochemistry and Biotechnology* (USA). 63/65: 449-456.

123. Zeng, Hong. Yan; Fang, Fang; Su , JieLong; Li, Chang Zhu and Jiang, LiJuan (2005). Technique of extracting oils from *Jatropha curcas* seeds. *Jiangsu Journal of Agricultural Sciences*. 21: 69-70.

124. Zou, J., Katavic, V., Giblin, E.M., Barton, D.L., MacKenzie, S.L., Keller, W.A., Hu, X. and Taylor, D.C. (1997) Modification of seed oil content and acyl composition in the Brassicaceae by expression of a yeast *Sn*-2 acyltransferase gene. *Plant* Cell 9: 909-923.

37. BIO-DIESEL: AT A GLANCE

ECO^3
BIO-DIESEL COULD BE BENIGN SOLUTION FOR ENERGY SECURITY, ECONOMY (RURAL) AND ENVIRONMENTAL CONSERVATION

ENERGY WILL BE GROWN BY OUR OWN FARMERS JUST AS THEY GROW FOOD CROPS

ENERGY SECURITY THROUGH ENERGY CROPS

GROW FUEL ENERGY FROM GREEN PLANTS

FARMERS CAN GROW ALL OUR FUEL AND FERTILIZER JUST AS THEY GROW ALL OUR FOOD FEED, FIBERS AND FIREWOOD

□□□